PhoneGap By Example

Use PhoneGap to apply web development skills and learn a variety of cross-platform mobile applications

Andrey Kovalenko

BIRMINGHAM - MUMBAI

PhoneGap By Example

First published: August 2015

Production reference: 1200815

Published by Packt Publishing Ltd.
Livery Place
35 Livery Street
Birmingham B3 2PB, UK.

ISBN 978-1-78528-531-8

www.packtpub.com

Credits

Author
Andrey Kovalenko

Reviewers
Rishi Bharat Jasapara

John Kershaw

Mohammad Nurdin bin Norazan

Commissioning Editor
Ashwin Nair

Acquisition Editor
Harsha Bharwani

Content Development Editor
Mamata Walkar

Technical Editor
Vijin Boricha

Copy Editors
Relin Hedly

Karuna Narayanan

Project Coordinator
Shipra Chawhan

Proofreader
Safis Editing

Indexer
Mariammal Chettiyar

Production Coordinator
Conidon Miranda

Cover Work
Conidon Miranda

About the Author

Andrey Kovalenko is a software developer, team leader, and blogger. He is a member of Jaybird Group, a web and mobile development firm in the U.S. and Ukraine. Andrey has been part of this organization since its inception and holds the position of a team leader. His job role involves overseeing and implementing projects in a wide variety of technologies with an emphasis on JavaScript, Node. js, HTML5, and Cordova (PhoneGap). Andrey leads several development groups that are responsible for creating products for call centers, marketing companies, real estate agencies, telecommunication companies, health care, and so on. Nowadays, he is focused on exploring the mobile development domain. As a result, Andrey started the BodyMotivator project, a mobile application for fitness. He believes that JavaScript has a great future as a generic development language. When Andrey isn't coding, he likes to spend time with his family and exercise at the local CrossFit gym. He is a health care enthusiast and is trying to use all his software development efforts to make his life healthier.

Andrey has authored *Instant KineticJS Starter*, *Packt Publishing*.

About the Reviewers

Rishi Bharat Jasapara started off his technical career as a Windows Phone developer and is still very much a Windows Phone loyalist. Along the way, he learned how to develop apps through PhoneGap and create a strong backend for products. Since he started his career in July 2012, he has worked with several start-ups. Rishi is currently the head of product at Timesaverz Dotcom Pvt. Ltd. Prior to this, he also worked as the chief technology officer for a start-up company based in New York and MobCast Innovations Pvt. Ltd., a Mumbai-based company. This is his first book as a reviewer. Rishi wishes to continue reviewing books for the younger, incoming breed of tech enthusiasts. You can find more information about him at `http://rishi.jasapara.me` and can connect with him on LinkedIn.

I would like to thank my parents, Bina and Bharat Jasapara, who continue to show immense faith in me and my decisions throughout. I would also like to thank my younger brother, Mohit, who has always been a bundle of youthful energy and encourages me to work hard and give my best. This book would not have been possible without the unconditional support that I received from Timesaverz Pvt. Ltd. I also want to acknowledge the contributions of my previous employers and colleagues, especially Ashwin Roy for being a brilliant mentor. I would also like to thank my close friends, relatives, teachers, and well-wishers, who have continued to support me throughout my life with their actions. Lastly, I would like to extend my gratitude to Packt Publishing for giving me the wonderful opportunity of reviewing this book.

John Kershaw has always had a fascination with how things work. This curiosity led him to pursue a master's degree in advanced computer science from the University of Manchester and move onwards to a career as a freelance software developer. John specializes in mobile apps and websites that are out of the ordinary.

He is also a freelance developer and the founder of Bristlr, a social network and dating site dedicated to people who love beards. When not running Bristlr, John gives talks on how to build start-ups, the joys of JavaScript, and how to be pragmatic while still using all the new buzzwords.

Mohammad Nurdin bin Norazan is a software engineer and a team leader with over 5 years of experience in mobile app development. He graduated with a bachelor's degree in information system engineering from Universiti Teknologi Mara, Malaysia. Mohammad is currently pursuing a master's degree in computer science from Universiti Teknologi Mara, Malaysia. His technical expertise includes frameworks, tools, and programming languages (including iOS, Android, PhoneGap, Cordova, IBM MobileFirst, Digital Ocean, Parse, and Heroku). Mohammad constantly delivers mobile app training, covering the iOS and Android platforms, to both students and professional audiences. He also works as a senior software engineer at Penril Datability (M) Sdn Bhd, a technical consultant at Technovault Solutions Sdn Bhd, a mobile technology team leader at Geomash/Dekatku Sdn Bhd, and a freelance programmer at Nurdin Norazan Services.

I would like to thank my parents, Norazan Zam and Mariah Hussin, and my beloved wife, Sabrina Hussin, for the immeasurable amount of support and guidance they have provided me. I would also like to thank my partner in crime, Asan Aldin, for his constant encouragement. I would like to take this opportunity to thank my former bosses, Lee Yong and Koh, for having faith in me when I worked with them. My sincere gratitude goes out to my teachers and lecturers, Dr. Mazlan, Saharbudin Naim, Dr. Suraya, and Dr. Nasiroh. Special thanks to my current bosses, Damien Santer, Graham Williams, and Tim Chandler, who trusted me and gave me the opportunity to lead the team. Last but not least, I would like to express my deepest gratitude to my family, friends, colleagues, and partners from the bottom of my heart for their help.

www.PacktPub.com

Support files, eBooks, discount offers, and more

For support files and downloads related to your book, please visit www.PacktPub.com.

Did you know that Packt offers eBook versions of every book published, with PDF and ePub files available? You can upgrade to the eBook version at www.PacktPub.com and as a print book customer, you are entitled to a discount on the eBook copy. Get in touch with us at service@packtpub.com for more details.

At www.PacktPub.com, you can also read a collection of free technical articles, sign up for a range of free newsletters and receive exclusive discounts and offers on Packt books and eBooks.

https://www2.packtpub.com/books/subscription/packtlib

Do you need instant solutions to your IT questions? PacktLib is Packt's online digital book library. Here, you can search, access, and read Packt's entire library of books.

Why subscribe?

- Fully searchable across every book published by Packt
- Copy and paste, print, and bookmark content
- On demand and accessible via a web browser

Free access for Packt account holders

If you have an account with Packt at www.PacktPub.com, you can use this to access PacktLib today and view 9 entirely free books. Simply use your login credentials for immediate access.

Table of Contents

Preface	**ix**
Chapter 1: Installing and Configuring PhoneGap	**1**
Downloading and installing	**2**
Installing Node.js on Mac	2
Installing Node.js from the official website	2
Installing Node.js with Homebrew	4
Installing Node.js on Windows	5
Installing Node.js on Linux	6
Installing PhoneGap with NPM	**7**
Understanding PhoneGap	**7**
Basic components	8
Development methods	8
Cordova installation	9
Creating an application	**10**
The config.xml structure	**12**
The iOS setup	**14**
Running the application in the iOS emulator	16
Running the application on an iOS device	**17**
Generating the iOS developer certificate	18
Adding the application identifier	20
Registering the device	22
Generating a provisioning profile	24
The Android setup	**27**
JDK Installation	28
Android SDK installation	28
Android Studio installation	30
Opening the project in Android Studio	33
Adding an Android emulator	34

PhoneGap development highlights	**37**
Use a single-page application approach	37
Don't generate the UI on the server	37
Limit network access	38
Increase perceived speed	38
Use hardware acceleration	38
Optimize images	38
Optimize payload	39
Minimize browser reflows	39
Test	39
Selecting a UI framework	**39**
Sencha Touch	41
jQuery Mobile	41
Ionic	41
Ratchet	41
The Kendo UI	42
Topcoat	42
React	42
Framework7	42
Famo.us	43
The Onsen UI	43
Summary	**44**
Chapter 2: Setting Up a Project Structure with Sencha Touch	**45**
An introduction to Sencha Touch	**46**
The installation of Sencha Touch	**47**
The installation of the Sencha Touch SDK	47
The installation of Sencha Cmd	47
Sencha Cmd features	**48**
Generating the application	**49**
Understanding the basic application structure	**54**
Getting familiar with the Sencha Touch view	56
Creating the Sencha Touch controller	58
Using store	60
The Sencha Touch model	61
The Sencha Touch store	62
The Sencha Touch proxy	64
Environment detection	64
Creating device profiles	64
Following the launch process	65
The UI and theming	66
Summary	**67**

Chapter 3: Easy Work with Device – Your First PhoneGap Application "Travelly"   **69**

Installing and using GapDebug   **70**
  iOS debugging setup   70
    Computer configuration   70
    iOS device configuration   71
  Android debugging setup   71
    Computer configuration   72
    Android device configuration   72
  The Genymotion Android emulator for faster debugging   75
The initial application's MVC structure   **75**
  Views   76
  Adding Pictos icons to the application   79
  Controllers   81
  Model and store   81
Using the Cordova StatusBar plugin to fix overlap   **82**
Using a camera to capture pictures   **85**
  Camera plugin installation   85
  Camera plugin usage   86
  Creating a new picture popup   88
Filesystem plugin installation and usage   **94**
  Using a persistent file location   94
Detecting the current geolocation   **97**
Saving data in local storage   **99**
Displaying data with Google Maps   **99**
  Displaying picture details in a popup   102
Summary   **106**

Chapter 4: Integrating the Travelly Application with Custom Service   **107**

Discovering the REST API   **108**
Exploring technologies to build a REST API   **109**
  Understanding Node.js   109
  Introducing MongoDB   110
  Installing MongoDB with Homebrew   110
Developing a REST API   **111**
  Using Express   111
  Generating an Express application   113
  Exploring the basic Express application   117
  Handling URLs with routes   119
  Returning a response   120
  Connecting Express and MongoDB   121

Creating a picture model 122
Creating a new picture record 125
Editing a picture record 125
Deleting a record 125
Implementing service authentication 126
Implementing a login form 127
Handling the authentication endpoint request 129
Verifying authentication 132
Implementing authentication on the application side **135**
Implementing file upload on the service side **140**
Implementing file upload on the application side **142**
Summary **144**
Chapter 5: Crazy Bubbles - Your First HTML5 Mobile Game **145**
What game framework to choose **146**
What is HTML5 Canvas? **146**
An introduction to Phaser **148**
Planning the game **149**
Generate a Cordova application **150**
Getting started with Phaser **151**
Download Phaser 151
Get tools 151
Use a web server 151
Prepare and create the game **153**
Preloading sprite **157**
Displaying sprite **158**
Handling pointer events with Phaser **160**
Handling the pointer move event **162**
Detect the bubble position under the pointer 163
Check whether a selected bubble can be moved to a new position 163
Swap bubbles 164
Releasing a bubble **166**
Check for matches 166
Remove matched bubbles 169
Drop down bubbles above the removed bubbles 169
Refill the board 170
Calculate score **172**
Running the application on the mobile **172**
Summary **173**

Chapter 6: Share Your Crazy Bubbles Game Result on Social Networks — 175

Implementing the game over screen — 175
The vertical scenario — 176
The horizontal scenario — 177
Coding the logic — 178
Implementing game restart — 184
Sharing on Twitter, Facebook, and other social media — 185
Sharing on Instagram — 189
Summary — 191

Chapter 7: Building a Real-time Communication Application – Pumpidu — 193

WebRTC fundamentals — 194
WebRTC audio and video engines — 195
The WebRTC protocol stack — 196
The RTCPeerConnection API — 197
The WebRTC browser support scorecard — 198
What is Crosswalk and why we need it? — 198
Adding Crosswalk support to the Cordova application — 199
Building our first real-time communication application — 200
Server side — 200
Client side — 202
Cordova application tweaks — 210
Running the application — 210
Building a real-time communication application with PeerJS — 214
Server side — 214
Client side — 216
Running the application — 221
Exploring other tools to build WebRTC mobile applications — 225
OpenTok — 225
PhoneRTC — 225
Summary — 226

Chapter 8: Building "Imaginary" – An Application with Instagram-like Image Filters — 227

An overview of the Pixastic library — 228
Bootstrapping the Sencha Touch application — 230
Capturing photos — 234

Rendering an effects list   236
  Including Pixastic   236
  showPhotoPopup   237
  Defining the effects model and store   241
  Applying effects to thumbnails   242
Applying effects to the photo   246
Saving the dressed photo into the application's folder   248
  Defining the picture model and store   248
  Saving the picture to the filesystem   249
Building a custom plugin to save the picture in the iOS library   252
  The plugin setup   253
  The JavaScript interface   254
  Native iOS code   255
  Publishing and using the plugin   257
Displaying the list of photos   258
Summary   262
Chapter 9: Testing the PhoneGap Application   263
Running with PhoneGap   263
  PhoneGap Developer App setup   264
  Handling code changes on the fly   266
  Including core plugins   266
Why we need tests   267
Testing theory   268
  Test-driven development   268
  Behavior-driven development   269
  Tests classification   270
    Unit testing   270
    Integration testing   270
    Functional testing   270
    System testing   270
    Performance or stress testing   270
Unit testing frameworks and test runners   271
Testing with Jasmine and headless browser PhantomJS   272
  Introduction to the Jasmine   272
    Writing unit tests with Jasmine   273
    Writing an integration test with Jasmine   277
  Writing Jasmine tests for Sencha Touch's Imaginary application   278
    Writing Jasmine tests for a controller   280
    Writing Jasmine tests for a model   282
  Running tests with the headless browser PhantomJS   284
Testing with DalekJS in a real browser   286

Performance testing with Appium and browser-perf	**289**
Other testing tools	**293**
Telerik Test Studio	294
Sauce Labs	294
Summary	**295**
Chapter 10: Releasing and Maintaining the Application	**297**
Versioning of the application	**298**
Using PhoneGap Build	**299**
PhoneGap config.xml	300
PhoneGap plugins	301
Initial upload and build	302
Beta release of the iOS application	**306**
Generate a distribution provisioning profile	307
Upload to iTunes Connect with Xcode	309
Upload to iTunes Connect with Application Loader	312
Invite internal and external testers	317
Release to the App Store	**319**
Release to Google Play	**322**
Create a keystore file	322
Build and sign an application in the release mode	322
Upload the application to the Google Play market	324
Using Fabric and Crashlytics	**325**
Summary	**334**
References	**334**
Index	**335**

Preface

PhoneGap is an open source framework that is responsible for creating mobile applications. The framework created by Nitobi Software. In 2011, Adobe purchased PhoneGap. You may have heard of Cordova. So, PhoneGap and Cordova are almost the same thing. Let's take a look at this in detail.

What this book covers

Chapter 1, *Installing and Configuring PhoneGap*, talks about the download and installation of the PhoneGap framework. It also examines the ins and outs of a basic PhoneGap application. In this chapter, you will learn how to perform basic manipulations with plugins and how to select the mobile web framework.

Chapter 2, *Setting Up a Project Structure with Sencha Touch*, explains the main elements of the framework. It also covers how to set maintainable, scalable, and testable project structures. It also teaches you how to follow the mobile-first approach and use CommonJS practices.

Chapter 3, *Easy Work with Device – Your First PhoneGap Application "Travelly"*, focuses on the application development tutorials with PhoneGap to build an application for travelers. You will learn how to access a camera to capture photo and how to work with the filesystem.

Chapter 4, *Integrating the Travelly Application with Custom Service*, adds new features to your applications, which allows you to sync data between the server and mobile device. You will learn how to build a custom web service with Node.js and integrate it with your PhoneGap application.

Chapter 5, Crazy Bubbles - Your First HTML5 Mobile Game, demonstrates the potential in building a HTML5 mobile game using the HTML5 Canvas and its 2D context. It teaches you how to build HTML5 animations, how to handle mobile gestures, and how to deal with performance issues.

Chapter 6, Share Your Crazy Bubbles Game Result on Social Networks, continues to provide information about the project from the previous chapter and adds the final touches to the game, including integration with the Facebook, Twitter.

Chapter 7, Building a Real-time Communication Application – Pumpidu, introduces all the popular WebRTC technologies and tells you how to build an audio/video chat with PhoneGap. It teaches you how to establish a video call among several mobile devices.

Chapter 8, Building "Imaginary" – An Application with Instagram-like Image Filters, shows you how to use the PhoneGap plugin in order to store pictures on the device. It also shows you how to implement Instagram-like picture filters.

Chapter 9, Testing the PhoneGap Application, teaches you how to use common approaches to test PhoneGap applications. You will learn how to use the key testing features on a real device and on a simulator.

Chapter 10, Releasing and Maintaining the Application, takes you through the process of how to release the application to different application markets, such as the App Store and Google Play. You will learn how to prepare the bundle for beta testing.

What you need for this book

In order to run the applications developed in this book, you will need Xcode and a Mac to run the PhoneGap application on iOS devices and Eclipse to run the PhoneGap application on Android devices. In order to publish an application to the App Store or Google Play, you will need accounts for these stores. I want you to pay attention to the fact that the developer account for publication in the App Store is a paid subscription. Also, we will be referring to the official PhoneGap documentation portal (`http://docs.phonegap.com/`) throughout .

Who this book is for

If you believe that mobile applications are the future of the information age and think that development should be quick and not a hustle, then this book is for you. You may be familiar with the fundamentals of JavaScript and HTML and have a basic understanding of cross-platform tools, but have no knowledge of PhoneGap. You may be interested in technologies or tools, such as Node.js, AngularJS, Grunt, Gulp, RequireJS, and so on. You will be able to build a real cross-platform mobile application in a short period of time and use the paradigm to build future projects. Building applications in this way will help you to shorten the release time.

Conventions

In this book, you will find a number of text styles that distinguish between different kinds of information. Here are some examples of these styles and an explanation of their meaning.

Code words in text, database table names, folder names, filenames, file extensions, pathnames, dummy URLs, user input, and Twitter handles are shown as follows: "At the end of the install, you will be prompted to make sure that /usr/local/bin is in your path."

A block of code is set as follows:

```
Ext.application({
    name: 'Travelly',
    views: [ 'Main' ],
    // ...
    launch: function() {
        Ext.fly('appLoadingIndicator').destroy();
        Ext.Viewport.add(Ext.create('Travelly.view.Main'));
    }
    // ...
});
```

When we wish to draw your attention to a particular part of a code block, the relevant lines or items are set in bold:

```
xtype: 'button',
text: 'My button',
id: 'myButton',
handler: function() {
    alert('My button has been clicked!');
}
```

Any command-line input or output is written as follows:

```
$ sencha generate model Picture id:int,url:string,title:string,lon:string
,lat:string
```

New terms and **important** words are shown in bold. Words that you see on the screen, for example, in menus or dialog boxes, appear in the text like this: "Once the installation is complete, you receive a **Successful Installation** message."

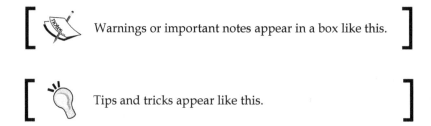

Warnings or important notes appear in a box like this.

Tips and tricks appear like this.

Reader feedback

Feedback from our readers is always welcome. Let us know what you think about this book—what you liked or disliked. Reader feedback is important for us as it helps us develop titles that you will really get the most out of.

To send us general feedback, simply e-mail feedback@packtpub.com, and mention the book's title in the subject of your message.

If there is a topic that you have expertise in and you are interested in either writing or contributing to a book, see our author guide at www.packtpub.com/authors.

Customer support

Now that you are the proud owner of a Packt book, we have a number of things to help you to get the most from your purchase.

Downloading the example code

You can download the example code files from your account at http://www.packtpub.com for all the Packt Publishing books you have purchased. If you purchased this book elsewhere, you can visit http://www.packtpub.com/support and register to have the files e-mailed directly to you.

Errata

Although we have taken every care to ensure the accuracy of our content, mistakes do happen. If you find a mistake in one of our books—maybe a mistake in the text or the code—we would be grateful if you could report this to us. By doing so, you can save other readers from frustration and help us improve subsequent versions of this book. If you find any errata, please report them by visiting http://www.packtpub. com/submit-errata, selecting your book, clicking on the **Errata Submission Form** link, and entering the details of your errata. Once your errata are verified, your submission will be accepted and the errata will be uploaded to our website or added to any list of existing errata under the Errata section of that title.

To view the previously submitted errata, go to https://www.packtpub.com/books/ content/support and enter the name of the book in the search field. The required information will appear under the **Errata** section.

Piracy

Piracy of copyrighted material on the Internet is an ongoing problem across all media. At Packt, we take the protection of our copyright and licenses very seriously. If you come across any illegal copies of our works in any form on the Internet, please provide us with the location address or website name immediately so that we can pursue a remedy.

Please contact us at copyright@packtpub.com with a link to the suspected pirated material.

We appreciate your help in protecting our authors and our ability to bring you valuable content.

Questions

If you have a problem with any aspect of this book, you can contact us at questions@packtpub.com, and we will do our best to address the problem.

1
Installing and Configuring PhoneGap

PhoneGap is an open source cross-platform framework used to build hybrid mobile applications. By hybrid, we mean HTML5, JavaScript, and CSS applications wrapped by a native shell. PhoneGap provides APIs to access the native function with JavaScript: accelerometer, camera, and so on.

All applications in this book have been developed on the Mac. However, you can easily carry out development on both Windows and Linux systems. The only small issue with PhoneGap development is building applications for Apple devices. If you want to test the application in an iOS emulator on your machine, you need a Mac operating system. If it is enough for you to build iOS applications with PhoneGap Build, then you could do it without Mac. However, you will need Mac in the initial stage to set up properly provisioning profiles and certificates for the iOS build with PhoneGap. You can use a friend's Mac terminal and then use Adobe PhoneGap Build to create an iOS rollout.

In this chapter, we will set up and configure everything on your computer so that you can develop and run all the applications in this book. This includes information about downloading, installation, and analysis of the basic PhoneGap application.

Also, this chapter is about a variety of mobile frameworks, comparison of them, how they fit with PhoneGap, and what framework or tool is better to choose. We will stick with a few mobile frameworks, but it is not compulsory for you to use this stack of approaches in your future projects. It is only a proposal for quick start. However, if you select some of my approaches in your real-world application, I will be more than happy.

This chapter will cover the following topics:

- How to install PhoneGap
- Understanding PhoneGap
- How to create a basic application
- How to configure an environment for iOS and Android development
- PhoneGap best practices
- What UI framework to select

Downloading and installing

When working with older versions of PhoneGap, we have to make a lot of detailed settings of the environment in order to run the application. However, with the newer versions, starting from 5.0.0, this procedure becomes easier. Before installing PhoneGap, we need to install Node.js, because it is easier to install PhoneGap CLI as a ready-to-use NPM package without compiling it from source codes. And NPM is a utility of Node.js.

 Node.js is a platform built on Chrome's JavaScript runtime. It was built as a tool for fast and scalable network applications. The main feature of the framework is an event-driven, non-blocking I/O model. For now, it is mainly used on server side in the same way as PHP, Ruby, or others. However, it is very popular and spreading fast nowadays.

There are several ways to install Node.js, but I will describe only two of them.

Installing Node.js on Mac

We will see how to install Node.js from the official website and with Homebrew.

Installing Node.js from the official website

To install Node.js, you can download a pre-compiled binary package, which makes for a nice and easy installation. Follow these steps:

1. Head over to `http://nodejs.org/` and click on the **INSTALL** button to download the latest package:

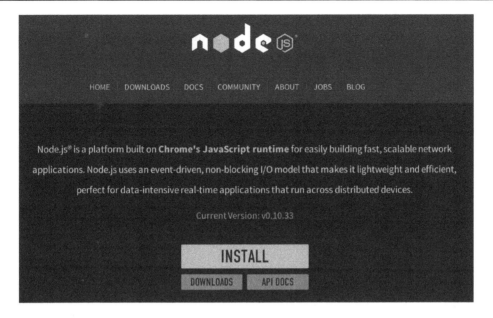

2. Install the package by following along. You will then be directed to install Node.js and **NPM** (**Node Package Manager**), which facilitates installs of additional packages for Node.js:

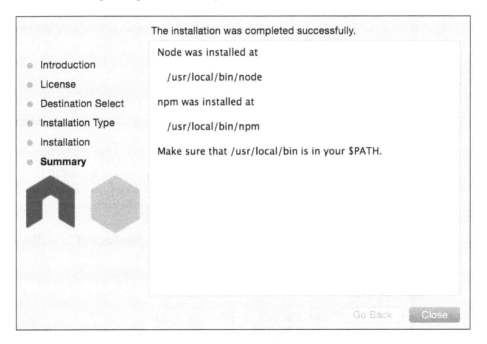

3. At the end of the install, you will be prompted to make sure that /usr/ local/bin is in your path. Double-check that you have it by running in the terminal using this command:

```
$ echo $PATH
```

If not, add it in either .bash_profile or .bashrc in your home directory.

4. After the installation, check whether it is OK by entering the following command in the command line node, which will open a Node.js JavaScript session:

```
$ node
> console.log('PhoneGap by Example');
PhoneGap by Example
undefined
```

5. To exit the Node.js session, just hit *control* + *c* twice. And we are done with the first method of Node.js installation.

Installing Node.js with Homebrew

Another good way to install Node.js is using Homebrew.

Homebrew (http://github.com/mxcl/homebrew) is the package manager that Apple forgot. Written in Ruby, it allows you to quickly and easily compile software on your Mac.

To install Homebrew (http://brew.sh/), follow these steps:

1. Run the following command on the console:

```
ruby -e "$(curl -fsSL https://raw.githubusercontent.com/Homebrew/
install/master/install)"
```

 The script explains what it will do and then pauses before it does it. It might require you to execute sudo and enter your root password. You should wait for a while to download and install all the components.

2. Once the installation is complete, you will receive a **Successful Installation** message:

```
==> Installation successful!
==> Next steps
Run `brew doctor` before you install anything
Run `brew help` to get started
```

3. Once Homebrew is installed, you can go ahead and install Node.js:

```
brew install node
```

It might require root access from you as well. And that is it. Node.js is installed now. It is pretty easy, right?

Installing Node.js on Windows

1. Download the installer from `https://nodejs.org/download/`.

2. Run the installer.

3. Follow the steps in the installer. One default option is to install NPM and another is to add Node.js to our path:

4. Test Node.js. Just open the Windows command prompt and type `node -v`. This should print a version number.

5. Test NPM. Type `npm -v` in the terminal. This should print NPM's version number.

Installing Node.js on Linux

To install Node.js on Linux, we should be familiar with the terminal as well. First of all, we need to install dependencies. Follow these steps:

1. Installing Ruby and GCC:

    ° For Ubuntu or Debian:

    ```
    sudo apt-get install build-essential curl git m4 ruby
    texinfo libbz2-dev libcurl4-openssl-dev libexpat-dev
    libncurses-dev zlib1g-dev
    ```

    ° For Fedora:

    ```
    sudo yum groupinstall 'Development Tools' && sudo yum
    install curl git m4 ruby texinfo bzip2-devel curl-devel
    expat-devel ncurses-devel zlib-devel
    ```

2. Installing Homebrew:

    ```
    ruby -e "$(curl -fsSL https://raw.githubusercontent.com/Homebrew/
    linuxbrew/go/install)"
    ```

After that, we need to add the following three lines to `.bashrc` or `.zshrc`:

```
export PATH="$HOME/.linuxbrew/bin:$PATH"
```

```
export MANPATH="$HOME/.linuxbrew/share/man:$MANPATH"
```

```
export INFOPATH="$HOME/.linuxbrew/share/info:$INFOPATH"
```

So, all the prerequisites are done, and we can install Node.js now. There are only two steps left to follow:

1. Open the terminal and type `brew install node`.
2. Wait until Homebrew finishes installation.

Now, we can test Node.js and NPM by running `node -v` and `npm -v` in the terminal accordingly.

Downloading the example code

You can download the example code fies from your account at http://www.packtpub.com for all the Packt Publishing books you have purchased. If you purchased this book elsewhere, you can visit http://www.packtpub.com/support and register to have the fies e-mailed directly to you.

Installing PhoneGap with NPM

We will use **NPM** (**Node Package Manager**) for all the future steps. NPM is part of Node.js, so we should already have installed it. Once you've installed Node.js, you can make sure you've got the most recent version of NPM using npm itself:

```
$ sudo npm install npm -g
```

(On Windows, you can drop sudo, but you should run it as administrator). Running this update will give you the most recent stable version of npm, also supported by NPM Inc.

So, we got Node.js and NPM installed. Let's install PhoneGap now. Open your command line and run the following command:

```
$ sudo npm install -g phonegap
```

Once the installation completes, you can invoke phonegap on the command line for further help. However, before that, let's understand how PhoneGap is organized.

Understanding PhoneGap

Further in this chapter, I will often mention **Apache Cordova** instead of PhoneGap, and we will use the Cordova command-line interface. This is considered to be more appropriate in the context of the mission of the library.

Let's add some clarity to the difference between these two technologies: Cordova and PhoneGap.

In a few words, PhoneGap is a distribution of Apache Cordova. PhoneGap can be considered as a shell for Cordova technology and provides great infrastructure for maintenance and distribution.

Cordova/PhoneGap includes native implementation for different mobile platforms. For example, on Android it is implemented with Java, and on iOS it is implemented with Objective-C.

In order to set up the PhoneGap project well, we should examine the basic concepts and principles of the application structure in detail.

Basic components

The app itself is an `index.html` web page by default, which connects a necessary CSS, JavaScript, images, media files, and other resources needed to run the application. The application runs in the native `WebView` wrapper, which can be spread through app stores.

The web view used by PhoneGap is the same web view used by the native operating system. On iOS, this is the Objective-C `UIWebView` class; on Android, this is `android.webkit.WebView`. There are differences in the web view rendering engines between operating systems, so we should account for this in our UI implementation.

The application can be fully or partially wrapped by WebView. For example, only some parts of the application may be made with HTML, and the remaining elements will be implemented with native components. In this book, we will consider that the applications are fully wrapped with WebView.

For interaction between our web page and the native components in Cordova, there is implemented plugins interface. This allows the JavaScript function to call the native components, and the native components transfer data in JavaScript. It might access the camera with JavaScript, accelerometer, or other device feature. Third-party plugins provide access to native features that are not necessarily present on every mobile platform. You can view all available plugins in the plugins repository (`http://plugins.cordova.io/`). You can also develop your own plugins. We will discuss this later.

Development methods

PhoneGap offers two approaches in the development of PhoneGap applications. These are cross-platform workflow and platform-centered workflow.

Cross-platform workflow: I recommend you to use this workflow if you intend to develop a mobile application under several platforms and you have no differences in the programs for various platforms. With this approach, Cordova CLI (The command-line interface) is primarily used. The Cordova CLI allows you to compile applications for different platforms, manage plugins, and so on.

Platform-centered workflow: I recommend you to use this workflow if you plan to focus on developing applications for a single platform and plan to quite deeply integrate with native components. This approach implies a certain development for a specific platform. For example, for iOS native development, you will use the Objective-C language, and for Android development, you will use the Java language.

PhoneGap allows you to move from a cross-platform workflow to a platform-centered workflow, but you cannot do it in the reverse order. The folder's structure is different and includes a different set of shell tools. Therefore, we start with the use of cross-platform workflow.

In this book, we will mostly use cross-platform workflow, but there will be times when you have to switch to platform-centered workflow. The difference will be in using platform-specific shell tools. For example, for iOS, we will run iOS-specific SDK shell commands. For example, for release, we will run this command:

```
$ /path/to/my_new_project/cordova/build -release
```

Cordova installation

In this section, we will discuss how to create an application and install it on different mobile platforms using the Apache Cordova command-line interface.

You will likely be surprised and ask about the difference between the PhoneGap command line and the Cordova command line, because we have already established PhoneGap. PhoneGap is a command-line utility that encapsulates Cordova. PhoneGap is built on Apache Cordova, and nothing else. You can think of Apache Cordova as the engine that powers PhoneGap. Over time, the PhoneGap distribution may contain additional tools, and that's why they differ in command, but they do the same thing. For the local build, PhoneGap uses the local library Cordova, but on the PhoneGap Build service, it uses it's own infrastructure. The official documentation for PhoneGap uses Cordova CLI.

 PhoneGap Build is an online service where we can build PhoneGap/Cordova application for distribution. In this case, we do not need to set up a build process locally. We will review this service in detail in *Chapter 10, Releasing and Maintaining the Application*.

We will also adhere to the official documentation necessary to carry out commands using Cordova CLI. Further, if necessary, we will use the PhoneGap CLI as well.

Before using the command-line tools, we needed to install the SDKs for each mobile platform we are targeting.

Cordova CLI supports the following combinations of platforms and operating systems:

Mobile OS	Windows terminal	Mac terminal	Linux terminal
iOS	-	+	-
Amazon Fire OS	+	+	+
Android	+	+	+
BlackBerry 10	+	+	+
Windows Phone 8	+	-	-
Windows	+	-	-
Firefox OS	+	+	+

On Mac, the command line is available via **Terminal Application**. On Windows PC, it's available as **Command Prompt** under **Accessories**.

We have already installed a PhoneGap library, but Cordova CLI also requires the installation.

For proper working of Cordova CLI, you must install a **Git** client, as the Cordova CLI refers to Git repositories to retrieve the necessary information. For more information on installing a Git client, you can refer to http://git-scm.com/. We should be able to run the Git command from the console. Once we have verified that the Git client is installed and running, let's install the Cordova CLI using NPM:

```
$ sudo npm install -g cordova
```

The preceding -g flag tells npm to install Cordova globally. Otherwise, it will be installed in the node_modules subfolder in the current folder.

Creating an application

Now, let's create our first application. Let's name it Travelly. It will be just a PhoneGap application scaffolding, and we will continue to develop it in the next chapter. So, to create an application, let's run the following command:

```
$ cordova create travelly com.cybind.travelly Travelly
```

We need to wait until this command is executed. The Cordova CLI refers to the external storage to extract all the files needed for the project:

- The first argument, `travelly`, is the folder where our project was generated.

- The second argument, `com.cybind.travelly`, provides our project with a reverse domain-style identifier. It is an optional argument, if we omit the third argument as well. We can edit this value later in `config.xml`. However, let's make it right from the beginning so that it is properly configured in the generated code as well. If we do not specify the identifier, it will be defaulted to `io.cordova.hellocordova`, which we do not want.

- The third argument, `Travelly`, provides the application's display title. It is an optional parameter, and the default value is `HelloCordova`.

Once the command execution is completed, a folder will appear with the following content:

```
$ ls
config.xml  hooks/     platforms/ plugins/   www/
```

Where:

- `config.xml`: This is the configuration file that contains important metadata needed to generate and distribute the application.

- `hooks/`: This is the folder for hooks and pieces of code that Cordova CLI executes at certain points in our Cordova application build.

- `platforms/`: This directory is for native code for each of the supported platforms. By default, this is empty, and we need to add the required platforms, which we will see later.

- `plugins/`: In this directory, we will place the plugins that provide extra support to the interface with each of the native platforms.

- `www/`: This directory houses our application's home page, along with various resources under `css`, `js`, and `img`, which follow common web development file-naming conventions. This gets copied into each of the platform-specific projects in platforms' folder whenever a build is performed. So, this is where our core development will happen and all our cross-platform code will live in.

You can see that all these folders are empty or contain just a basic set of files. The `config.xml` file contains minimum information. The folder will get full, and `config.xml` will grow as we continue our development. So, get ready.

The config.xml structure

Before proceeding to consider the specific settings for each mobile platform, let's look at the common configuration file `config.xml`. This file contains very important information on setting up our future applications.

By default, our `config.xml` file has the following contents:

```xml
<?xml version='1.0' encoding='utf-8'?>
<widget id="com.cybind.travelly" version="0.0.1" xmlns="http://www.
w3.org/ns/widgets" xmlns:cdv="http://cordova.apache.org/ns/1.0">
    <name>Travelly</name>
    <description>
      A sample Apache Cordova application that responds to the
      deviceready event.
    </description>
    <author email="dev@cordova.apache.org"
    href="http://cordova.io">
      Apache Cordova Team
    </author>
    <content src="index.html" />
    <access origin="*" />
</widget>
```

Where:

- **Widget**: It's ID attribute provides the app's reverse-domain identifier, and the version provides its full version number.

 Reverse domain name notation is a naming convention for the components, packages, and types used by a programming language, system, or framework.

- **Name**: This specifies the app's formal name as it appears on the device's home screen and within app-store interfaces.

- **Description and author**: This specifies metadata and contact information that may appear within app-store listings.

- **Content**: This optional element defines the app's starting page in the top-level web assets directory.

- **Access**: This defines the set of external domains the app is allowed to communicate with. In our case, we allow it to access any server.

Furthermore, there can be other options presented. Usually, they are added under the tag access. These elements are preference and feature.

Preference items can be global and multiplatform.

For example, the following two settings are global and apply to all supported platforms:

```
<preference name="Fullscreen" value="true" />
<preference name="Orientation" value="landscape" />
```

Where:

- Fullscreen allows you to hide the status bar at the top of the screen
- Orientation allows you to lock orientation and prevent the interface from rotating in response to changes in orientation

The following two settings apply to multiple platforms, but not all:

```
<preference name="TopActivityIndicator" value="gray" />
<preference name="AutoHideSplashScreen" value="false" />
```

Where:

- TopActivityIndicator sets the color of the Activity Indicator
- AutoHideSplashScreen specifies whether to hide the splash screen automatically or allow the programmer to do it in code

In this case, it is not necessary to add feature elements manually, because in the initial stage, we will use the cross-platform workflow, where we will use the command CLI plugin to add the device API. However, when we move to fine-tuning of each platform, we will add the feature elements, as shown in this example:

```
<feature name="Device">
  <param name="ios-package" value="CDVDevice" />
</feature>

<feature name="Device">
  <param name="android-package"
  value="org.apache.cordova.device.Device" />
</feature>
```

The iOS setup

In order to be able to run the application being developed in an iOS simulator or on an iOS device connected to our computer, we need the following components:

- OS: Mac OS X

- IDE: Xcode (6.0 and newer)

- iOS SDK

 You can download Xcode from `https://developer.apple.com/xcode/downloads/` and iOS SDK from `https://developer.apple.com/ios/download/`.

The only disappointment when developing for iOS with the ability to debug on your computer is a limitation of the operating system by Apple. Unfortunately, it must only be Mac OS X operating system that does this. If you do not want the ability to run applications on your computer, you can simply use the service PhoneGap Build. However, as we try to better understand the features of PhoneGap, we should look deeper into platform-specific aspects.

We can test many of the Cordova features using the iOS emulator installed with the iOS SDK and Xcode, but we need an actual device to fully test all of the app's device features before submitting it to the App Store. To install apps onto a device, we should be a member of Apple's iOS Developer Program, which costs $ 99 per year. Next, we will describe how to run our application in an iOS emulator, which does not require the acquisition of the program.

So, let's say, we already have Mac OS X installed. The next thing we need to do is install Xcode. It is very simple. Follow these steps:

1. By keyword "Xcode", find the application in the App Store and press the **Install App** button. Once Xcode is installed, several command-line tools need to be enabled for Cordova to run.

2. From the Xcode menu, select **Preferences**.

3. Then, click on the **Downloads** tab.

4. From the **Components** panel, press the **Install** button next to the command-line tools listing.

5. On the same window, you can install other components, such as several versions of an iOS simulator.

Now, with the help of the Cordova CLI, we can add our future version of iOS applications:

```
$ cd travelly
$ cordova platform add ios
Creating ios project...
```

Now, in the platform folder, the `ios` subfolder appeared, and in the plugins folder, the `ios.json` file appeared. Let's open `travelly/platforms/ios/Travelly.xcodeproj` in Xcode.

The Xcode window should look as follows:

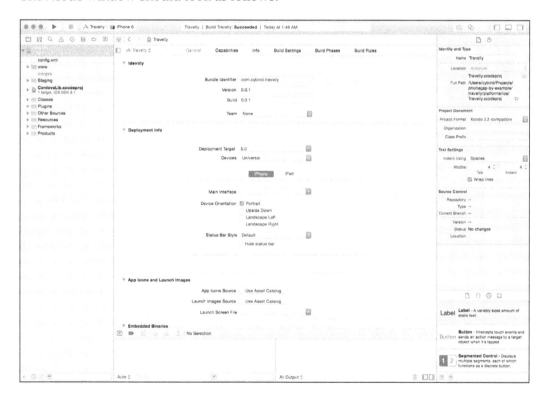

Running the application in the iOS emulator

Now, let's run our application in the iOS emulator using the following steps:

1. Select the intended device from the toolbar's **Scheme** menu, such as the **iPhone 6**, as highlighted here:

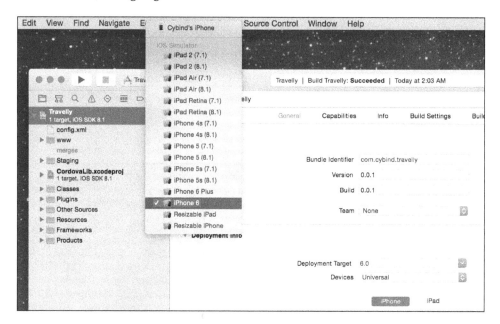

2. Press the **Run** button that appears in the same toolbar to the left of **Scheme**. This button builds, deploys, and runs the application in the emulator. A separate emulator application opens to display the app:

A similar procedure can be done with the help of the Cordova CLI:

```
$ cordova build ios
```

This generates `ios` platform-specific code within the project's platforms subdirectory.

The `cordova build` command is a shorthand for the following command:

```
$ cordova prepare ios
$ cordova compile ios
```

To run our application in the iOS emulator, it is enough to execute the following command:

```
$ cordova emulate ios
```

We will see the same application in the emulator that we saw when run from Xcode.

Running the application on an iOS device

To run the application on an iOS device, you must perform the following steps:

1. Join the Apple iOS Developer Program.
2. Create a Provisioning Profile within the iOS Provisioning Portal.
3. Verify that the Code Signing section's Code Signing Identity within the project settings is set to our provisioning profile name.

Let's assume that we already have a subscription to the iOS Developer Program. Now, you need to create an iOS Development Certificate and Provisioning Profile.

Go to `https://developer.apple.com`, then go to the "**Member Center**" link, and enter your Apple ID and password. After that, go to **Developer Program Resources** and select **Certificates, Identifiers & Profiles**. The following screen will appear:

Generating the iOS developer certificate

Under the **Certificates** section, press the **+** button in order to start certificate generation. On the screen, select **iOS App Development**.

Then, press the **Continue** button. After that, we see the instructions for how to generate **Certificate Signing Request (CSR)** on our computer. At this stage, we can begin to generate the CSR on our computer. In the `Applications` folder on your Mac, open the `Utilities` folder and launch **Keychain Access**. Within the **Keychain Access** drop-down menu, go to **Keychain Access | Certificate Assistant | Request a Certificate from a Certificate Authority**. After that, we will see the following screen:

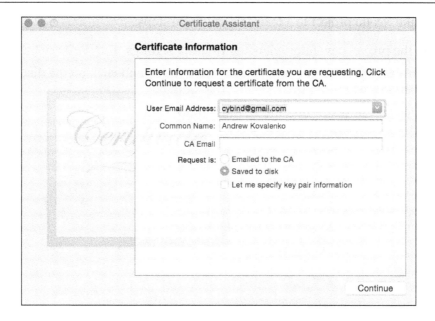

In the **User Email Address** field, I entered my current e-mail. Usually, this is the same e-mail to access the Apple Developer Portal. In the **Common Name** field, I usually enter my name. If you have a company, it might be the company name. I am the owner of an Apple Developer account, so I do not need to send a request by e-mail. I just clicked on **Save to disk** and then on the **Continue** button. I saved the CSR file on your computer and completed the process of generating the Certificate Signing request.

Now, go back to the browser with instructions for the generation of CSR and click on **Continue**. I selected the `.certSigningRequest` file saved on my Mac and clicked on the **Generate** button. Once the certificate generation is over, I downloaded a ready certificate on my computer and double-clicked on it to add to my Mac keychain.

At this stage, the generation of the certificate is completed. Now, you need to attach your project team. The Xcode project needs to be assigned to a team so that Xcode knows where to create your code signing and provisioning assets.

To assign the Xcode project to a team, follow these steps:

1. In the project navigator, view the **Identity** settings.

2. If necessary, select the target, click on **General**, and click on the disclosure triangle next to **Identity** to reveal the settings.

3. Choose your team from the **Team** pop-up menu.

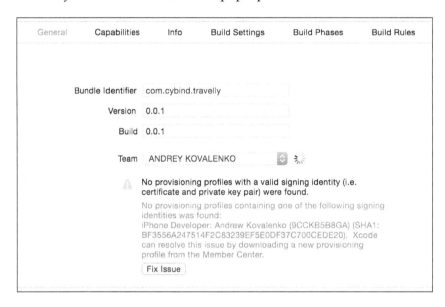

This message appears because we have not generated a Provisioning Profile yet.

Adding the application identifier

In order to generate a Provisioning Profile for our application, it must be registered with the Apple Developer Portal. To do this, on the presented Xcode screen, copy Bundle Identifier to the clipboard and switch to Apple Developer Portal. In the **Identifiers** section, press the + button. On the next page, you must enter the application **Name** and paste **Bundle ID** from the clipboard, as presented in the following screenshot:

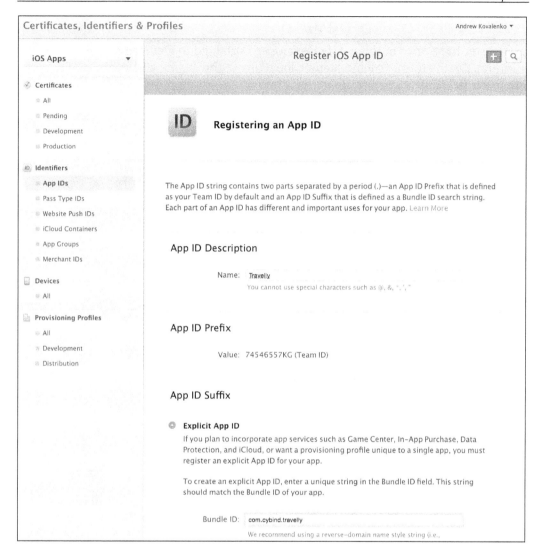

After that, click on the **Continue** button, and on the next page, check the entered data and click on **Submit**. We added our application. Now, we need to register the device where we will be able to run our application.

Registering the device

To register the device in the Apple Developer Portal, go to **Devices** and click on the + button. On the form, you must enter the name of the device and its **unique device identifier** (UDID).

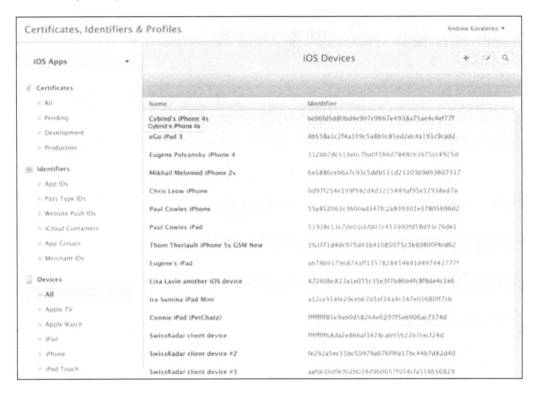

UDID can be found by connecting the Apple device to your computer in iTunes. If you go to the **Device Manager** tab and then to **Summary**, you will see the following screen:

Here, in the main section, you can copy the UDID by right-clicking on the **UDID** value. After that, go to the Apple Developer Portal, insert **UDID**, enter the **Name** of your device, and click on the **Continue** button:

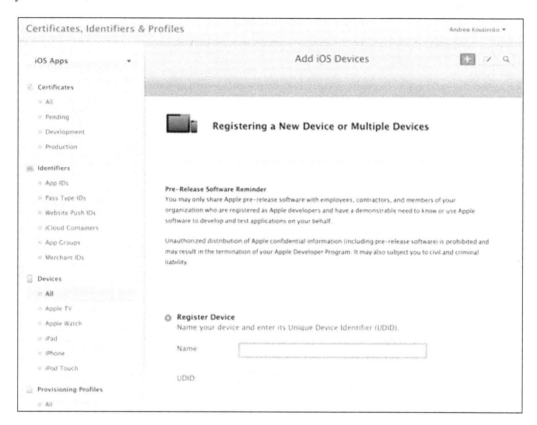

Once the operation is complete, we will see our device in the **Devices** section. The next step is Provisioning Profile generation.

Generating a provisioning profile

To generate a Provisioning Profile for our application, go to the **Provisioning Profiles | Development** and press the **+** button. After that, select the **iOS Development** application type and click on **Continue**.

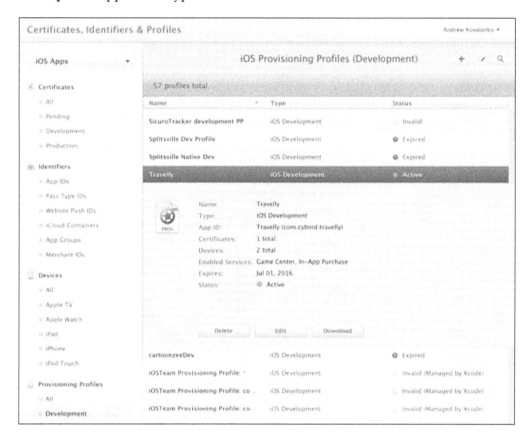

In the next step, we will choose our application, **Travelly**, and go to the next page. Select the certificate generated by us and move on to the next step. In this step, we will select the devices on which we want to run our application:

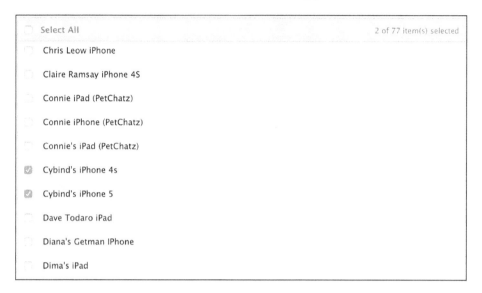

At this stage, I chose two devices. I want to be able to test our application on both devices.

Finally, go to the next page and enter a name of the Provisioning Profile. Let's enter **Travelly**. Complete the generation by pressing the **Generate** button. Then, we can download this Provisioning Profile, or allow Xcode to it.

Now, in Xcode, we need to select a proper Provisioning Profile. Follow these steps:

1. In the project navigator, view the **Identity** settings.
2. If necessary, select the target, click on **Build Settings**, and click on the disclosure triangle next to **Code Signing** to reveal the settings.

3. Select **Travelly** as **Provisioning Profile**.

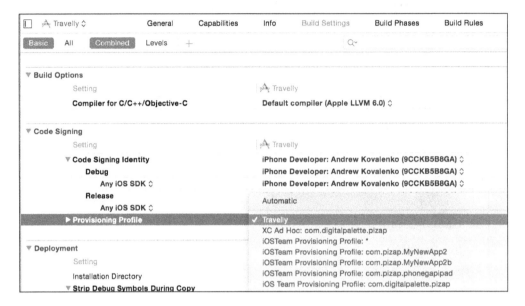

Alternatively, you can simply click on a provisioning profile, and it will be automatically installed in the Xcode.

That's all. Now, you can run the application on the device. Connect your device to your computer using the USB cable. Select the intended device from the toolbar's scheme menu; in my case, it is **Cybind's iPhone**. Press the **Run** button that appears in the same toolbar to the left of the scheme. This button builds, deploys, and runs the application in the device. We can see how the application started on our device.

Similarly, we can run the application on our device from the console:

```
$ cordova run ios
```

This command displays a lot of information because the Cordova CLI builds iOS projects using Xcode, which displays all the information about how it builds the application.

Congratulations! We have successfully executed our initial application on the real iOS device.

The Android setup

Setting up the project to run the application on the Android platform looks a little easier, but also has its complexities. To run the application in the Android simulator or on an Android device connected to our computer, we need the following components:

- OS: Linux or Windows or Mac
- Java: Oracle JDK
- IDE: Android Studio
- Android SDK

First, let's look at what platforms we have already added. This can be done using the platform list command of the Cordova CLI:

```
$ cordova platform list
Installed platforms: ios 3.7.0
Available platforms: amazon-fireos, android, blackberry10, browser,
firefoxos
```

As you can see, we have only added the `ios` platform. Now, let's try to add the Android version of our application using the command platform, `add`:

```
$ cordova platform add android
```

If we had never been developing for Android on this computer, we could get an unsuccessful response. This can be as follows:

```
[Error: The command 'android' failed. Make sure you have the latest
Android SDK installed, and the 'android' command (inside the tools/
folder) added to your path.
```

It can also look like this:

```
Error: ANDROID_HOME is not set and "android" command not in your PATH.
You must fulfill at least one of these conditions.
```

It only means that we need to configure our environment for Android development. For a quick development start, Google has prepared the Android Studio. It includes the essential Android SDK components and IDE. Let's get started with JDK, Android SDK and Android Studio installation.

JDK Installation

So, I do not have Java installed. I go to the official website, `http://www.oracle.com/technetwork/java/javase/downloads/index.html`, to get the latest version of it. I download JDK for MAC OS X and install it. You should pay attention that it should be JDK, not JRE. Only JDK provides the required functionality to build our Android application. After JDK installation let's run this command and check whether Java has been successfully installed:

```
$ java -version
java version "1.8.0_25"
Java(TM) SE Runtime Environment (build 1.8.0_25-b17)
Java HotSpot(TM) 64-Bit Server VM (build 25.25-b02, mixed mode)
```

OK! We have successfully installed the JDK.

Android SDK installation

Let's try to add the Android platform in our application using the Cordova CLI again:

```
$ cordova platform add android
Creating android project...
Error: Please install Android target "android-19".
Hint: Run "android" from your command-line to open the SDK manager.
```

Apparently, we have not installed all the necessary packages for development under Android 4.4.2 (API 19) yet. Open the Android SDK Manager using the `android` console command. Select **Android 4.4.2 (API 19)** and press **Install**. The installation process should look like this:

Android Wear Intel x86 Atom System Image	20	3	Not installed
Sources for Android SDK	20	1	Not installed
Android 4.4.2 (API 19)			
SDK Platform	19	4	Not installed
Samples for SDK	19	6	Not installed
ARM EABI v7a System Image	19	2	Not installed
Intel x86 Atom System Image	19	2	Not installed
Google APIs (x86 System Image)	19	9	Not installed
Google APIs (ARM System Image)	19	9	Not installed
Glass Development Kit Preview	19	11	Not installed
Sources for Android SDK	19	2	Not installed
Android 4.3 (API 18)			
Android 4.2.2 (API 17)			

You also need to install all the following packages in order to make sure that they all run smoothly:

- Android SDK Tools
- Android SDK Platform Tools
- Android SDK Build Tools
- Google USB Driver
- Intel x86 Emulator Accelerator (HAXM installed)

For Cordova command-line tools to work, or the CLI that is based upon them, we need to include the SDK's tools and platform-tools directories in our path. On a Mac, I use a text editor to create or modify the `~/.bash_profile` file, adding a line such as the following one:

```
export PATH=${PATH}:/Development/adt-bundle/sdk/platform-tools:/
Development/adt-bundle/sdk/tools
```

You can see the path to your Android SDK in the SDK Manager window. You can launch it by going to **Android Studio | Tools | Android | SDK Manager**.

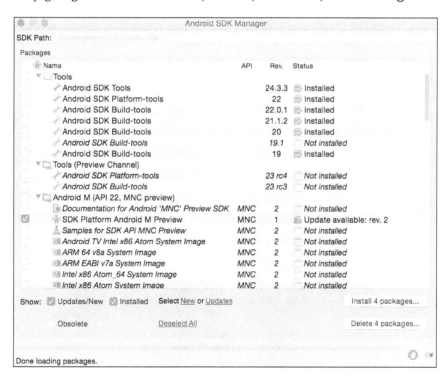

Here, in the top-left corner, you can see the **SDK Path** value. It is what we included in the path.

Once all the required packages are installed let's try to add the Android platform again:

```
$ cordova platform add android
Creating android project...
Creating Cordova project for the Android platform:
  Path: platforms/android
  Package: com.cybind.travelly
  Name: Travelly
  Android target: android-19
Copying template files...
Project successfully created.
```

And we have successfully done it! Now, in our platforms directory, we got the android subfolder. In the plugins folder, there appeared a new file named android.json.

Android Studio installation

To get started, go to https://developer.android.com/sdk/index.html and download **Android Studio** (with the Android SDK for Mac):

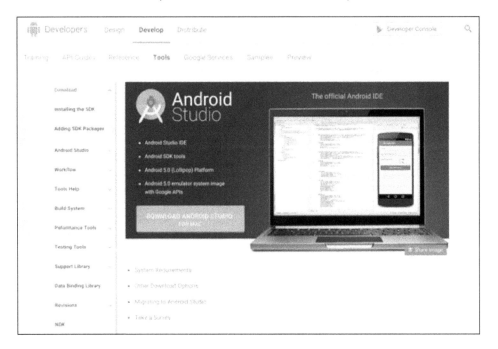

Once downloaded, install it using the following steps:

1. Launch the `.dmg` file.
2. Drag and drop Android Studio into the `application` folder.
3. Open Android Studio and follow the setup wizard.

There will be one step where we need to add a path to the JDK. In my case, it looks like this:

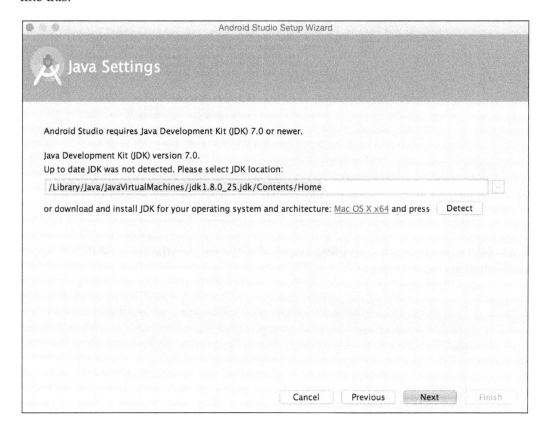

On the license agreement screen, we will accept all the points:

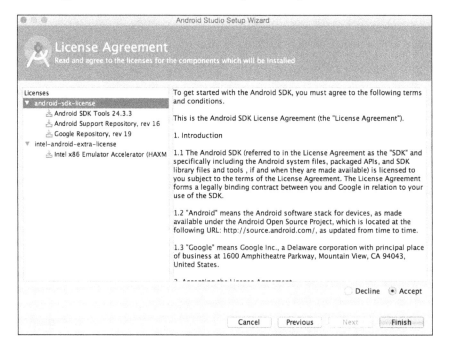

It will take some time after that, and eventually, we will see the successful installation screen:

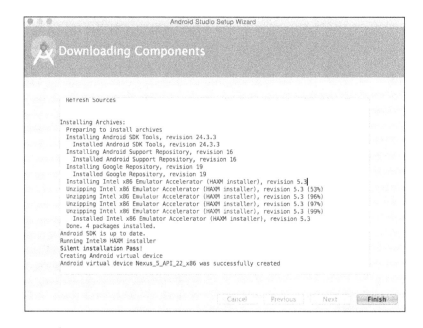

Opening the project in Android Studio

Now, let's try to open our Android project in Android Studio by following these steps:

1. Launch the Android Studio application.

2. Select **Import project (Eclipse ADT, Gradle, etc)**:

3. Select the location where the Android platform is stored:

4. Press **OK**.

Now, we can build and run the Android application directly from Android Studio.

We can add the Android emulator device now.

Adding an Android emulator

Open the terminal and enter the following command:

```
$ android avd
```

In the opened window, click on **Create**. In the window that appears, enter the parameters of the emulator. In my case, it looks like this:

> Be sure to select **Intel Atom (x86)** in **CPU/ABI**. We need it to enable hardware-accelerated emulation. Sometimes, developers choose **ARM** by mistake that can cause the simulator to run slowly.

I entered the name of the device, selected the **Nexus 4 (4.7", 768 x 1280: xhdpi)** device, and targeted **Android 4.4.2 - API Level 19**.

After that, I clicked on the **OK** button. We successfully added the emulator.

Go back to the main window of Eclipse, highlight the project, and click on the **Run** button. As a result, we see our project running in the Android emulator:

So, we can run our application in the emulator from the console with this command:

```
$ cordova emulate android
```

To run the application on the device, we only need to connect it to our computer and run this command:

```
$ cordova run android
```

That's all! We have completed the configuration of our project to run on iOS and Android emulators and devices. Now, let's discuss some of the best development practices for PhoneGap applications.

PhoneGap development highlights

We should mention some highlights before starting real development. They will help you understand why some mobile frameworks behave the way they do. The first thing we should understand is that a mobile device is more limited in resources than a computer. We should think of it from different aspects of the development.

Use a single-page application approach

Single-page application (SPA) is a web application or website that fits on a single web page with the goal of providing a more fluid user experience.

Loosely defined, a SPA is a client-side application that is run from one request of a web page. The user loads an initial set of resources (HTML, CSS, and JavaScript) and further updates (showing a new view, loading data) are done via AJAX.

Some SPA libraries you can use in your Cordova applications are:

* AngularJS
* EmberJS
* Backbone
* Kendo UI
* Monaca
* ReactJS
* Sencha Touch
* jQuery Mobile

Don't generate the UI on the server

Often, we need an interaction between our application and other servers. We need it to share data between multiple endpoints: other mobile devices, websites, and so on. Very often, the architecture is built in such a way that it sends not just data, but layout information as well.

It will be better to just create a needed set of data on the server side and send it with JSON, XML, or other formats. It can be customized as well. Do not send HTML through the Internet because it could be stored on the client. We will reduce the payload without sending HTML through Internet.

Limit network access

Cache static and dynamic data on the device. It can be filesystem, local storage, or database. Use the offline-first approach. We will discuss this approach in the upcoming chapters.

Increase perceived speed

We can create the illusion of faster hybrid application with the following approaches:

- Don't wait for the data to display the UI

 Do not show the preloaders without it being ready UI. Display the UI first, and only when you get data, update this UI. It allows you to increase perceptive performance.

- Avoid the click event's 300 ms delay

 Do not use click events on the mobile devices. It works fine on the devices, but most devices impose a 300 ms delay on them in order to distinguish between a touch and a touch hold event. Using touchstart or touchend will result in a dramatic improvement—300 ms doesn't sound like much, but it can result in jerky UI updates and behavior.

Use hardware acceleration

Using hardware-accelerated CSS transitions will be dramatically better than using JavaScript to create animations. See the list of resources at the end of this section for examples.

Optimize images

Combine images in sprites. It will decrease the number of requests and will improve the speed of image display. Just use CSS sprite sheets, which support high-resolution screens.

You can read about sprites on the following sites:

- `https://en.wikipedia.org/wiki/Sprite_(computer_graphics)`
- `http://webdesign.tutsplus.com/articles/css-sprite-sheets-best-practices-tools-and-helpful-applications--webdesign-8340`

There are a lot of other resources that you can find online as well.

Optimize payload

Compress CSS and JavaScript. Compress JPEG pictures. Don't use a full-stack framework just because you like a small piece of it. Use system fonts. Use fonts for icons.

[

For example, we can use FontAwesome from `http://fortawesome.github.io/Font-Awesome/`. It includes a lot of free icons we can use.
]

Minimize browser reflows

Minimizing browser reflows will help in saving memory and CPU resources. We can do it with following steps:

- Reduce the number of DOM elements
- Minimize access to the DOM
- Update elements "offline" before reinserting into DOM
- Avoid tweaking layout in JavaScript

Test

Use Chrome Developer tools, Xcode profiler, and other tools to understand performance problems, memory leaks, and other issues in the application.

Selecting a UI framework

When you start building your application, you should think about the user, and how users of the specific platform feel about it: they just want the app to behave as expected. Your first stop should be each platform's design guidelines. Mobile OS manufacturers typically have official docs geared for professionals developing for their platform. These docs often include guides on UI design patterns.

- For Android, check out Android designs (`http://developer.android.com/design/patterns/index.html`) pattern section
- For iOS app designers, see iOS Human Interface Guidelines (`https://developer.apple.com/library/ios/documentation/UserExperience/Conceptual/MobileHIG/index.html`) in the iOS Developer Library

Implementing all of these components, patterns, and animations on your own can be quite a challenge. That is why there are already implemented frameworks with different meanings. Some are good in UX, some are sloppy. Some are light, some are heavy.

UX: This stands for user experience design. It is a process of enhancing customer satisfaction by improving usability, simplicity, and pleasure of the product.

UI: This stands for user interface. It is mainly a set of interface elements for presentation of the UX.

A great product experience starts with UX followed by UI.

When choosing a framework, in any technology — whether it is your frontend or backend — it is always important to remember what you are trying to achieve.

Are you building an iOS-only application or is it cross-platform? Are you trying to make impressive animations and transitions and smooth UX? Do you want to deploy your product fast? Does your target audience have high-end mobile devices, or are they mostly using old phones?

However, I think, what really matters is performance.

Here is short list of UI frameworks I know about:

- Sencha Touch
- jQuery Mobile
- Ionic
- Ratchet
- Kendo UI
- Topcoat
- ReactJS
- Framework7
- Famo.us
- Onsen UI

Let's look briefly at each framework.

Sencha Touch

This framework is pure JavaScript and CSS. Sencha Touch is a mature framework built to fit the most demanding app needs. It is a versatile enterprise-level workhorse using CSS3 and HTML5 best practices. It uses the Model–ViewController–Store pattern meant for serious business, with over 50 built-in components, from the basic ones such as component, container, form, and various fields to more complex carousel, lists, pickers, charts, grid, and much more. Sencha Touch includes themes for every popular mobile platform. Sencha Touch abstracts device APIs for their native packager and Cordova/PhoneGap. It has a large community and very detailed documentation. It takes some time to learn, but the documentation is awesome and everything goes smoothly. Unfortunately, it has a commercial license.

jQuery Mobile

This jQuery-based framework is the most commonly used mobile application HTML5 framework. This is because jQuery is popular and transition to jQuery mobile is very easy. However, it is sluggish on mobile devices; it is not optimized as is Sencha Touch. It has an average UI, and the official documentation is lacking some information. There is no MVC support. It is even more sluggish when combined with PhoneGap.

Ionic

Ionic is an open source framework. It is a pretty new library. Since its release, it has gained a lot of respect in the hybrid and mobile development community. Ionic provides a lot of utilities, and it is pretty easy to do small customizations for iOS, Android, and other operating systems. It uses Cordova for packaging from the box.

Ratchet

This is an open source framework from the Twitter Bootstrap creators. It provides a basic UI for iOS and Android (it's just UI components). To organize your application properly, you should architect it properly or use an additional library to do it for you. It means it doesn't support any kind of MV* pattern.

The Kendo UI

This jQuery-based framework is beautiful. It supports MVVM and has its own support for server-side communication (.NET, PHP and Java). It will cost you some money if you want to build a commercial application. It has great template support; every template looks like a native template. It supports MVC from the box and has great documentation.

Topcoat

This is just an open source library with well-done CSS. When you develop with Topcoat, you should care about the application structure, JavaScript, and performance on your own. It doesn't provide application wire framing or scaffolding.

React

This is the only view framework from Facebook. Yes, the developers had a little crooked soul; they did not make it for the sake of marketing. In fact, React is a view-oriented MVC framework, although it does not appear as such at first sight. The biggest advantage of React is that is uses a virtual DOM diff implementation for ultra-high performance. It is great with performance on the mobile device, but the learning curve is steep, and the documentation is still not too detailed or understood.

Framework7

This is a free and full-featured HTML framework for building iOS apps. It is also an indispensable prototyping apps tool to show a working app prototype as soon as possible, in case you need to. It is easy to use and customize. It provides its own MVC framework. However, I would not use it for multiplatform development, because it represents only iOS UX/UI.

Famo.us

The Famo.us's mobile framework is a newborn baby. It's the newest framework on the market nowadays. Famo.us targets a specific need at this time: performance. With 60 fps animations, Famo.us is the choice for you if you want to brag in your hybrid app. Basically, Famo.us uses its own JavaScript engine that works with the GPU acceleration provided by CSS3 3D transformation functions to make animations as smooth as can be at 60 fps. Of course, we will not be able to reach 60 fps performance on old devices. This is for devices with good graphic cards. The Famo.us team is trying to tie it closely with the AngularJS framework. It is a good choice when building complex user experience animations, but the learning curve is steep, and the documentation is still not too detailed or understood.

The Onsen UI

Onsen UI is an Ionic competitor. It also comes with AngularJS support and provides the same solution as the Ionic team meant to build. Onsen UI was born as an answer for PhoneGap and Cordova developers who were struggling with the UI when starting a project, as the Internet lacks in mobile UI frameworks. The documentation in some places is not clear. I do not really like that I cannot use the default HTML layout but their own with `ons-` tags.

When I was researching these libraries, I compared them long enough to understand what I should choose to implement applications in this book. The main problem was the choice between Sencha Touch, Ionic and Onsen UI. Only these libraries have a sufficient set of ready-to-use components for the iOS and Android platforms and have solutions for an MV* pattern and a fairly good performance level on mobile devices. After some deliberation, I decided to choose the library that is the oldest on the market, has a huge community, and has good documentation. So, I selected Sencha Touch.

Summary

You just learned how to install and configure PhoneGap for iOS and Android development. You also learned how to create a basic application and successfully run it on emulators and real mobile devices. We discussed main frameworks to help us in future application implementation.

Keep the project structure as we developed here, and we will extend and improve it in the next chapter. You will learn how to organize your HTML/CSS/JavaScript part of the Cordova/PhoneGap application in the most graceful way.

2
Setting Up a Project Structure with Sencha Touch

The biggest misunderstanding with PhoneGap is that it does everything for you.

However, that's not the case. It is just the base for your application. It helps you package your app and access device features, such as the camera. There is nothing in PhoneGap that helps you to organize your app in terms of, for example, MVC. It's not an application framework. You need more. You need help from the PhoneGap ecosystem. By ecosystem, I mean everything that helps us build mobile-ready websites.

You need to spend a lot of time building an app that feels and looks native. It's the details that eat up your budget. Again, it's not PhoneGap's fault. Surely, JavaScript will perform faster on more modern devices soon. Surely, PhoneGap will remain a great service. However, we now need other great tools that can be used with PhoneGap to make HTML5/JS on mobile devices a success.

In this chapter, we will build a foundation of our Travelly application using PhoneGap and Sencha Touch.

This chapter will cover the following topics:

- Sencha Touch and the issues it solves
- Offline first approach
- Installing Sencha Touch
- Installing the Sencha Touch command-line tool
- Setting up a project via the command-line tool
- The Model–View–Controller–Store pattern

An introduction to Sencha Touch

In the previous chapter, I decided to select Sencha Touch as a framework for application development. It has been in existence for some years now and is popular among hybrid mobile application developers.

Sencha Touch is a product of Sencha company. It was first released on July 17, 2010. It was the 0.90 beta version. It was formed after other popular frameworks, such as Ext JS, jQTouch, and Raphael. At the time of writing this book, the latest version was 2.4.2.

It allows you to develop mobile applications that would have the same look and feel as a native application. Sencha Touch supports Android, iOS, Windows Phone, Microsoft Surface Pro and RT, and Blackberry devices.

Getting started with Sencha Touch isn't that difficult, but in order to get the best out of Sencha Touch, we need to invest a considerable amount of time in it.

To get a feel of a Sencha Touch app, take a look at the samples provided on its official page at `http://www.sencha.com/products/touch/demos/`.

You can also access its documentation on the official website at `http://docs.sencha.com/touch/`.

The main challenge in building a hybrid mobile application is performance optimization. In the *PhoneGap development highlights* section of *Chapter 1, Installing and Configuring PhoneGap*, we already mentioned the best practices to follow so that the application performs well. Sencha Touch helps us follow these practices by:

- Organizing a single-page application
- Creating the UI in JavaScript at the client side
- Inserting/removing views into/from the DOM as needed
- Avoiding network access
- Caching static data
- Caching dynamic data
- Using CSS transitions and hardware acceleration
- Using native scroll
- Avoiding click event's 300 ms delay

We will look at these approaches once we install Sencha Touch.

The installation of Sencha Touch

Download the free Sencha Touch SDK (`http://www.sencha.com/products/touch/download/`) and Sencha Cmdr. (`http://www.sencha.com/products/sencha-cmd/download`) from the Sencha website.

The installation of the Sencha Touch SDK

Extract the SDK zip file somewhere you usually place your SDKs. In my case, it is `/Development/touch-2.4.1`. Now, we will use this folder in the future for our project-generation process.

The installation of Sencha Cmd

To get Sencha Cmd working, you should install Java Runtime Environment (`http://www.oracle.com/technetwork/java/javase/downloads/index.html`), Ruby, Apple Xcode for iOS packaging, Android SDK, and Android Studio for Android packaging. We got all this packages installed, except Ruby.

Not all OSes have Ruby installed. Here are the OS-specific instructions to download Ruby:

- **Windows**: Download Ruby from `rubyinstaller.org`. Get the `.exe` file version of the software and install it.

- **Mac OS**: Ruby is preinstalled. You can test whether Ruby is installed with the `Ruby -v` command.

- **Ubuntu**: Use `sudo apt-get install ruby2.0.0` to download Ruby.

So now, everything is ready for the Sencha Cmd installation. Extract the Cmd zip file into the `temporary` folder. Run the Sencha Cmd installer. All the steps in the Sencha Cmd installation are clear without any input of data, so you just need to click on the **Next** button several times. The installer adds the Sencha command-line tool to your path, enabling you to generate a fresh application template, among other things. In my case, it has added the following row to the `~/.bashrc` file:

```
export PATH=/Users/cybind/bin/Sencha/Cmd/5.1.0.26:$PATH
```

 At the time of writing the book, I used the latest available Sencha Touch SDK and Cmd packages. I got Sencha Touch SDK 2.4.1 and Sencha Cmd 5.1.0.26.

To verify that Sencha Cmd installed successfully, simply type the `sencha` command in the terminal. You should see output that starts like this:

```
$ sencha
Sencha Cmd v5.1.0.26

...
```

Before continuing to the application generation, we should better understand what we can use from Sencha Cmd and how it can help us.

Sencha Cmd features

The Sencha command-line tool helps us with different aspects in application development.

Sencha Cmd parameters are organized in categories and commands. Let's review some of them that we will use.

Here are the categories that we will use:

- **App**: This performs various application build processes
- **Generate**: This generates models, controllers, stores, or an entire application
- **Web**: We will use this to test our application in a browser; it manages a simple HTTP file server
- **Compass**: This wraps the execution of compass for Sass compilation
- **Compile**: This compiles sources to produce concatenated output
- **Cordova**: This manages Cmd/Cordova integration
- **PhoneGap**: This manages Cmd/PhoneGap integration

Here are the commands that we will use:

- `help`: This displays help for commands
- `upgrade`: This upgrades the Sencha Cmd

Generating the application

Let's try some of the Sencha Cmd features and generate the Travelly application. First of all, we should change our current directory to Sencha Touch SDK and run the Sencha generate command there:

```
$ cd /Development/touch-2.4.1
$ sencha generate app Travelly ~/Projects/phonegap-by-example/sencha-travelly
```

The same result can be achieved without changing the directory to the SDK, but passing an additional argument within the command:

```
$ sencha -sdk /Development/touch-2.4.1 generate app Travelly ~/Projects/phonegap-by-example/sencha-travelly
```

Where:

- `/Development/touch-2.4.1` is the directory where you unzipped the Touch SDK
- The `generate` command is used to generate the appropriate part of the application; in our case, it is entire application
- `Travelly` is the name of our application
- `~/Projects/phonegap-by-example/sencha-travelly` is the path where we want our application to be generated

There is a new `sencha-travelly` folder created with the following file structure:

```
├── app
├── app.js
├── app.json
├── bootstrap.js
├── bootstrap.json
├── build
├── build.xml
├── index.html
├── packages
├── resources
└── touch
```

Where:

- `index.html`: This is the page from where our application will be hosted.
- `app`: This is the main application directory. In general, it is a collection of Models, Views, Controllers, Stores, and Profiles, which are explained here:
    - **Model**: This represents the type of data that should be used/stored in the application
    - **View**: This displays data to the user with the help of inbuilt Sencha UI components/custom components
    - **Controller**: This handles UI interactions and the interaction between the Model and the View
    - **Store**: This is responsible for loading data into the application
    - **Profile**: This helps in customizing the UI for various phone and tablets
- `app.js`: This is the main JavaScript entry point for our app. It contains the following elements:
    - The app name, and references to all the models, views, controllers, profiles, and stores.
    - The app launch function that is called after the models, views, controllers, profiles, and stores are loaded. The app launch function is the starting point of the application, where the first view gets instantiated and loaded.
- `app.json`: This is the configuration file for our app.
- `bootstrap.js`: This file should *not* be edited. It is provided to support global includes and other.
- `bootstrap.json`: This file should *not* be edited. It is a combination of contents from `app.json` and all the required package's `package.json` files.
- `build`: This is a folder where Sencha Touch places the built application.
- `build.xml`: This is the configuration file for the build process.
- `packages`: This folder serves as the storage of all packages used by the applications (or other packages) in the workspace. We will not work with packages for now.
- `resources`: This is a directory, and it contains images, CSS, and other media assets.
- `touch`: This is a directory, and it contains the Sencha Touch framework files.

You might be interested in why we generated the application under the `sencha-travelly` folder while we already have the `travelly` folder with a configured PhoneGap project structure. Sencha Touch already has a good integration with PhoneGap and with Sencha Cmd. So, we will explore it on a new, clear project.

We can easily do this with the following command:

```
$ sencha cordova init com.cybind.travelly Travelly
```

Where:

- `com.cybind.travelly`: This is a bundle identifier of our application. Remember that we entered the same in Xcode in *Chapter 1, Installing and Configuring PhoneGap*.
- `Travelly`: This is the name of our application for our packaged application.

You will notice that in root folder of our application, we got the `cordova` subfolder with the following content:

```
├── config.xml
├── hooks
│   └── README.md
├── platforms
├── plugins
└── www
    ├── css
    ├── img
    ├── index.html
    └── js
```

It is exactly the same basic Cordova application that we generated in *Chapter 1, Installing and Configuring PhoneGap*. Only our `platforms` folder is still empty. Let's fix it.

Just open the `app.json` file in a text editor, find section builds in it, and uncomment the row with platforms list. So, your builds section will look something like this:

```
"builds": {
    "web": {"default": true},
    "native": {
        "packager": "cordova",
        "cordova" : {
            "config": {
                "platforms": "ios android",
```

```
                    "id": "com.cybind.travelly",
                    "name": "Travelly"
                }
            }
        }
    }
```

Here, you can see that by default, the build for Web is turned on, and as a packager, Sencha Touch uses Cordova. Also, in the `config` section, you can see the `id` and `name` properties, which we got from our `cordova init` command.

Now, we can run our application in iOS and Android emulators simply by executing this command:

```
$ sencha app build -run native
```

In the preceding command, we specified that we should build our application first and run the native application only after this. Here is a step-by-step working of the command:

1. Build the Sencha Touch web application.

2. Get information about the available platforms from the preceding `config` section.

3. Generate projects for these platforms.

4. Copy the built web application into the platforms' web folders:

    ◦ `cordova/platforms/ios/www` for iOS

    ◦ `cordova/platforms/android/assets/www` for Android

5. Start the default native emulator or device for each platform; we already configured them in *Chapter 1, Installing and Configuring PhoneGap*.

6. Deploy our Sencha Touch application onto every running device.

7. Run our application.

As a result, we will see something similar to this screenshot:

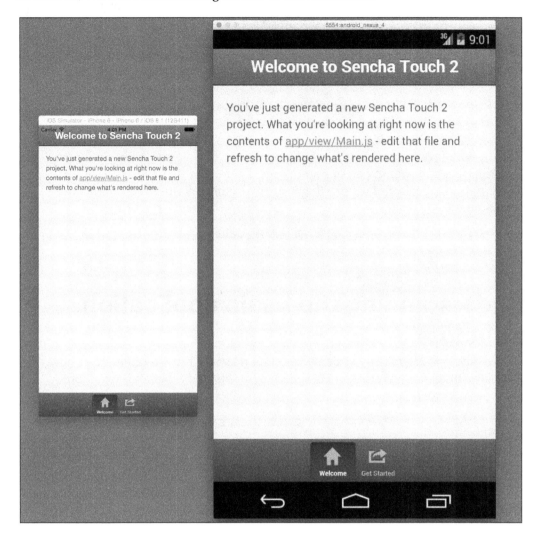

You might get the following error when running an iOS simulator:

`Error: ios-sim was not found. Please download, build, and install version 3.0 or greater from https:// github.com/phonegap/ios-sim into your path. Or npm install -g ios-sim using Node.js: http://nodejs.org/.`

In this case, we need to install the `ios-sim` package, as presented earlier. Just run the `npm install -g ios-sim` command in the terminal. Repeat the build command again.

This is kind of cool, right? Not much development, but we got the basic Sencha Touch application running on our emulators. I agree that there are a lot of configuration aspects, but it's worth getting everything configured for effective development.

Now, let's go deeper into the Sencha Touch framework and prepare the web side of our application.

Understanding the basic application structure

We already looked at the filesystem structure. It is now time to understand how these files are tied together. You also need to understand what an interaction between code parts is.

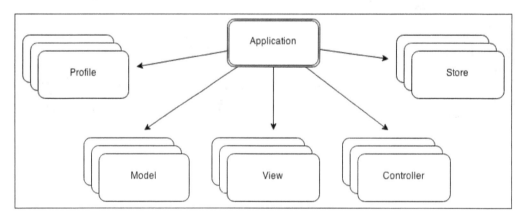

`Ext.application` is the starting point in our application. As we noted earlier, it might contain the app name, and references to all the models, views, controllers, profiles, and stores. These are explained as follows:

- **Profiles**: These allow us to customize the application's UI for handsets and tablets

- **Models**: These represent a type of data in our application

- **Views**: These actually present data in our application within Sencha Touch components

- **Controllers**: These handle interactions with our application by listening for user's taps and swipes

- **Stores**: These store our data, which we display in grids and other elements

You can see the single instance of `Ext.application` in the generated `www/app.js` file:

```
Ext.application({
    name: 'Travelly',
    views: [ 'Main' ],
    // ...
    launch: function() {
      Ext.fly('appLoadingIndicator').destroy();
      Ext.Viewport.add(Ext.create('Travelly.view.Main'));
    }
    // ...
});
```

Where:

- **Travelly**: This is the name of our generated application. We will use it as a global namespace in different places of our application, for example, `Travelly.controllers.Main`, `Travelly.model.Main`, `Travelly.view.Main`.

- **Main**: This is the name of our one and only view in the application.

- **Launch**: This is a method called once the application loader has loaded all the required dependencies.

- **.fly**: This is Sencha's method used to make a one-time reference to DOM. It is taking HTML element on the `index.html` page with `id="appLoadingIndicator"` and destroys it.

- **Viewport.add**: This inserts `Main` view into `index.html` body.

Sencha Touch is built using lessons learned from Ext JS (http://www.sencha.com/products/extjs). That is why, in our application, you can see the Ext prefix in many places.

The current file structure inside the app folder looks like this:

```
├── controller
├── form
├── model
├── profile
├── store
└── view
    └── Main.js
```

It simply becomes clear that in the controller folder, we should place our controllers, in view our views, and so on. Now, all our folders, except view, are empty. Let's fix it and create several items we will need for our application.

Getting familiar with the Sencha Touch view

Our application has already a defined view in the view/Main.js file:

```
Ext.define('Travelly.view.Main', {
    extend: 'Ext.tab.Panel',
    xtype: 'main',
    requires: [ 'Ext.TitleBar', 'Ext.Button' ],
    config: {
      tabBarPosition: 'bottom',
      items: [
        {
          title: 'Welcome',
          iconCls: 'home'
          // ...
        },
        {
          title: 'Get Started',
          iconCls: 'action'
          // ...
        }
      ]
    }
});
```

Let's break the preceding code down.

The Ext.define function helps us define a class named Main in the Travelly/view namespace. All view components are placed within this namespace as per Sencha Touch MVC standards.

The extend keyword specifies that the Main class is the subclass of Ext.tab.Panel. So, the Main class inherits the base configuration and implementation of the Ext.tab.Panel class.

The xtype keyword is used to instantiate the class.

The requires property is used because we use a button in our items array. We indicate the new view to require the Ext.Button class. At the moment, the dynamic loading system does not recognize classes specified by xtype, so we need to define the dependency manually.

The Config keyword helps us initialize the variables/components used in that particular class. In this example, we should initialize the tab panel view with tabs and panels. Tab panels are a great way to allow the user to switch between several pages that are all full screen.

The content of Items in the Main view currently has two items with title and iconCls. The tabBarPosition property defines where the tab bar is placed, and in our case, it is at the bottom. With title, we can set text on a button in the tab panel, and with iconCls, we can show the predefined icon. Let's add a button to one of this tab panels.

To add a new element to any container (in our case, it is tab panel), it is enough to assign an array of objects to items property. Let's do this with our first tab panel:

```
{
    title: 'Welcome',
    iconCls: 'home',

    items: [
        {
            xtype: 'button',
            text: 'My button',
            id: 'myButton'
        }
    ]
}
```

You can see that we added some text and assigned `id` for the button. This will help us handle button events in the controller. However, we can already handle it right in the view. Here is an example:

```
xtype: 'button',
text: 'My button',
id: 'myButton',
handler: function() {
    alert('My button has been clicked!');
}
```

Once the user has tapped on the button, we will see an alert message right away.

And we are done! We just successfully added the Sencha Touch button component to the first tab of our application. We should remember that our application already has the `Ext.Viewport.add(Ext.create('Travelly.view.Main'))` code to add our view on the page.

Now, let's create some controller and handle user interaction on the `Main` view.

Creating the Sencha Touch controller

Controllers listen for events fired by the UI and take actions based on the event. Using controllers helps keep your code clean and readable, and separates the view logic from the control logic.

There are two ways to create the controller: using Sencha Cmd or manually. To create the controller with Sencha Cmd, it is enough to execute the following command:

```
$ sencha generate controller Main
```

It generates the `controllers/Main.js` file for us. It is generic and has minimum content:

```
Ext.define('Travelly.controller.Main', {
    extend: 'Ext.app.Controller',
    config: {
      refs: {},
      control: {}
    },
    launch: function(app) {}
});
```

Our controller is a subclass of Ext.app.Controller, which is instantiated only once by the application that loaded it. At any time, there is only one instance of each controller. To instantiate the controller automatically, we have to add it into the controllers configuration section in the application in the following way:

```
controllers: [ 'Main' ]
```

You can see the launch method. This method is triggered automatically for every controller, every time the application starts. If you do not need it, simply remove this method from the controller.

The config section is different from the view's section, but it contains two important properties: refs and controls.

The refs property is an easy way to find components on the page. The control property is similar to the ref's config property, but it allows us to define event handlers. Here is an example:

```
config: {
    refs: {
        myButton: '#myButton'
    },
    control: {
        myButton: {
            tap: 'doMyButtonTap'
        }
    }
},
doMyButtonTap: function() {
    alert('My button has been clicked!');
}
```

In the refs section, we are using ComponentQuery to find our button on the page by id. Whenever a button of this type fires its tap event, our controller's doMyButtonTap function is called.

This is mostly what a controller does – listens for events that fire (usually by the UI) and initiates some action.

 Ext.ComponentQuery lets us search of components within Ext.ComponentManager (globally) or a specific Ext.Container on the document with syntax similar to a CSS selector.

Using store

When web developers think about storing something for the user, they try to upload data to the server and store it there. However, it is a huge issue in mobile web and hybrid applications development when you do not have stable Internet connection. Sometimes, a device can go offline and go online only in several days. We have to stop building apps with a desktop mindset where we have permanent, fast connectivity. Offline technologies don't just give us sites that work offline; they improve performance and security by minimizing the need for cookies, HTTP, and file uploads. They also opens up new possibilities for better user experience.

HTML5 allows us to make the approach real. Now, it is possible to store data on client side in WebStorage, IndexedDB, and Web SQL Database, which are explained here.

- **LocalStorage**: This is also known as web storage, simple storage, or by its alternate session storage interface. This API provides synchronous key/value pair storage.
- **Web SQL Database**: This offers more full-featured database tables accessed via SQL queries.
- **IndexedDB**: This offers more features than LocalStorage, but fewer than Web SQL.

However, there is a limitation for using these client-side storages with the PhoneGap application. Here is a list of mobile platforms supporting different storages:

- LocalStorage
    - All supported by PhoneGap platforms
- WebSQL
    - Android
    - BlackBerry 10
    - iOS
    - Tizen
- IndexedDB
    - BlackBerry 10
    - Windows Phone 8
    - Windows 8

You can learn more about browser storage support at `http://www.html5rocks.com/en/tutorials/offline/quota-research/`.

Now, it is clear what storages we can use with PhoneGap. It is great news that Sencha Touch already has an abstraction for the storage — `Ext.data.Store`.

Before looking deeper into `Ext.data.Store`, let's get familiar with Sencha Touch models. These models are very important, because they are the main components of the store.

The Sencha Touch model

This model represents objects of business models from our application. For example, we want to store metadata for pictures we capture with camera. In our Travelly application, we can define `model Picture`. Let's generate it as we did for controller:

```
$ sencha generate model Picture id:int,url:string,title:string,lon:string
,lat:string
```

Where:

- `model`: This is an attribute for the `generate` command to generate the model
- `Picture`: This is a name of our model to generate
- `id:int,url:string,title:string,lon:string,lat:strin`: These are fields for our model with their types

In the `model` folder, we should see the `Picture.js` file:

```
Ext.define('Travelly.model.Picture', {
    extend: 'Ext.data.Model',
    config: {
      fields: [
        { name: 'id', type: 'int' },
        { name: 'url', type: 'string' },
        { name: 'title', type: 'string' },
        { name: 'lon', type: 'string' },
        { name: 'lat', type: 'string' }
      ]
    }
});
```

Where I have used the following components in the file:

- `id` as a unique identifier for the picture
- `url` as a link to the picture
- `title` of the picture
- `lon` as a longitude coordinate of the place where the picture was taken
- `lat` as a latitude coordinate of the place where the picture was taken

To work with this model in our application, we should add `Travelly.model.Picture` to the `requires` section. Usually, we do this in controllers or stores. To create an instance of this class, use the following lines of code:

```
var picture = Ext.create('Travelly.model.Picture', {
    id: 1,
    url: 'http://myurl.my/somepicture.jpg',
    title: 'My picture',
    lat: '50.450783',
    lon: '30.523035'
});
```

Where:

- `Ext.create`: This instantiates a class by either full name, alias, or alternate name
- `Travelly.model.Picture`: This is the name of our model class

We can access model object values by property names. For example, `picture.get('title')` will return the title of our picture.

The Sencha Touch store

Let's create a data store and map it to the preceding model. There is no Sencha Cmd command for it. So, we will do this manually:

```
Ext.define('Travelly.store.Pictures',{
    extend:'Ext.data.Store',
    config:{
      model:'Travelly.model.Picture',
      autoLoad:true,
      data:[
        {
          id: 1,
```

```
            url: 'http://myurl.my/somepicture1.jpg',
            title: 'My picture 1',
            lat: '50.450783',
            lon: '30.523035'
            },
            {
              id: 2,
              url: 'http://myurl.my/somepicture2.jpg',
              title: 'My picture 2',
              lat: '50.450783',
              lon: '30.523035'
            }
          ],
          proxy:{
          type:'localstorage'
        }
      }
    });
```

Where:

- `autoLoad: true`: This means the store's load method is automatically called after creation. We do not need to call the `.load` method before working with data.

- `data`: This is a inline data based on our model.

- `proxy`: This is used to load and save data. We used `LocalStorage`.

The Sencha store provides an easy way to add and retrieve data from the store:

```
var pictureStore = Ext.getStore('Pictures');
pictureStore.add(picture);
pictureStore.sync();
var foundPic = pictureStore.findRecord('id', 1);
```

Where:

- `Ext.getStore` gets the instance of the `Pictures` store that is already defined and initialized

- `pictureStore.add` adds our early created picture object to the store

- `pictureStore.sync` syncs our in-memory store data to LocalStorage

- `pictureStore.findRecord` finds and retrieves the `Picture` model instance from the in-memory storage

The Sencha Touch proxy

There are two types of proxies that could be used: client and server.

These are the client proxies:

- `LocalStorageProxy`: This stores data in LocalStorage
- `MemoryProxy`: This stores data in memory only

These are the server proxies:

- `Ajax`: This is used to interact with a server on the same domain
- `JsonP`: This uses JSONP to send requests to a server on a different domain

We will use server proxies when we integrate our application with RESTful service on Node.js.

Environment detection

Very often, when we run the application, we need to know on which platform we are running it. With Sencha Touch, we can detect this information:

- **Operating system**: `Ext.os.is.[iOS, Android, MacOS, Windows]`, and so on
- **Device**: `Ext.os.is.[iPhone, iPad, iPod]` and so on
- **Browser**: `Ext.browser.is.[Safari, Chrome, Firefox]` and so on.
- **Browser's features**: `Ext.feature.has.[Canvas, Css3dTransforms]` and so on.

In our case, it is very helpful to use environment detection when we develop. We can detect whether it's desktop browser or mobile device, and run or avoid device-specific functions. Alternatively, we can run different code for Android and iOS.

Creating device profiles

Very often, you need to make the application behave differently on different devices. We can separate code execution by form factor or by operating system. It doesn't matter which we use.

Let's imagine that we need some specific functionality running only on Android tablet, and on all the other devices, we would like to keep it generic.

To generate a specific profile, let's execute the Sencha Cmd command:

```
$ sencha generate profile AndroidTablet
```

It generates the `profile/AndroidTablet.js` file for us. It is pretty basic, and I modified it to look like this:

```
Ext.define('Travelly.profile.AndroidTablet', {
    extend: 'Ext.app.Profile',
    config: {
      views: [],
      models: [],
      stores: [],
      controllers: [ 'Main' ]
    },
    isActive: function(app) {
      return Ext.os.deviceType == 'Tablet' && Ext.os.is.Android;
    }
});
```

Once the profiles have been loaded, their `isActive` functions are called in turn. If these functions return true, the application loads all of its dependencies—the models, views, controllers, and other classes.

Following the launch process

Each application, profile, or controller can define the launch function. However, they don't have to.

Here is the order of functions called after the application starts:

1. Controller's init
2. Profile's launch
3. Application's launch
4. Controller's launch

[

Only the active profile has its launch function called.
]

The `init` method is called by the controller's application to initialize the controller. Here, we can place any pre-launch logic. At this stage, we do not have UI ready.

By the time the chain of calls reaches the controller's launch, we have already prepared the UI, because mainly, all UIs are created by the profile's or application's launch functions.

The UI and theming

While Sencha Touch initially favored an iOS-styled interface, its main theme is not platform oriented and does not mimic any of the mobile OSes entirely. Nonetheless, iOS, Android, and Windows lookalike themes are shipped with the entire package. Sencha Touch utilizes the SASS approach to its fullest extent. We can quickly restyle a look and feel for our needs. I do not really like the default Sencha Touch theme. So, let's try to change it to iOS cupertino.

Let's do this using these steps:

1. Run the `compass watch` command. Sencha Cmd does it for us, with the following command:

   ```
   $ sencha app watch
   ```

2. Open up the `/resources/sass/app.scss` file in the text editor

3. Save a copy of the `app.scss` file and rename it to `cupertino.scss`

4. Modify `cupertino.scss` so that it looks like this:

   ```
   @import 'sencha-touch/cupertino';
   @import 'sencha-touch/cupertino/all';
   ```

5. Link our newly generated `cupertino.css` file in the `app.json` file. Let's make the following changes to it:

   ```
   "css": [
       {
           "path": "resources/css/cupertino.css",
           "update": "delta"
       }
   ]
   ```

And we are done! When we started the `$ sencha app watch` command, you might have realized that it showed the `localhost` URL address to test application in output. In my case, it was presented in the following way:

```
$ sencha web start
Sencha Cmd v5.1.0.26
[INF]  Mapping http://localhost:1841/ to ....
[INF]  ---------------------------------------------------------------
[INF]  Starting web server at : http://localhost:1841
[INF]  ---------------------------------------------------------------
```

So I opened the `http://localhost:1841` URL in the browser and was able to see the application. Similarly, you can use the `$ sencha web start` command. However, it is not watching for `scss` changes. In the upcoming chapters, we will often use the simple Sencha web server for our testing and debugging needs.

Summary

Sencha Touch plays well with PhoneGap, mainly because they are both created to complement each other. Sencha Touch handles the web structural and UI parts, and PhoneGap handles the infrastructure to run on mobile devices. However, what we found out is that Sencha Touch is build to support the developer. It removes all those boring tasks, such as DOM element referencing, attaching event listeners, and so on. It also provides some ready-to-use components to work with LocalStorage. It's a great choice for the modern mobile hybrid application coding stack.

In the next chapter, you will learn how to access a device's native functions through the plugin interface. You will better understand the main feature of the PhoneGap application and learn how it functions as a bridge between the web and native parts.

3
Easy Work with Device – Your First PhoneGap Application "Travelly"

In this chapter, we will continue to build a mobile application "Travelly" using PhoneGap, its plugins, and Sencha Touch. We will utilize the application structure already defined in the previous chapter. It is a mobile application for a traveller to capture pictures and place them on the map. Our system should support the addition, display, and removal of media elements. Our application should be able to access hardware features of the mobile device it is running on. It will access the device's camera to capture photos, the filesystem to save media data in persistent file storage, and geolocation to detect the place of the captured photo. We will use Sencha Touch to build the user interface and control the client-side logic. The Sencha Touch store will handle the storage of information, and Google map will be used to display saved media.

This chapter will show you how to:

- Use GapDebug
- Install PhoneGap plugins
- Access the camera to capture a photo and save it to the device's filesystem
- Detect the current device's geolocation
- Use the Sencha Touch store
- Display data on Google Maps

Installing and using GapDebug

Developers have used PhoneGap (Cordova) for hybrid mobile application development since 2009. The PhoneGap application expanded to include iOS, Android, and many other platforms. However, while the wrapper itself is great, debugging tools, unfortunately, are not ideal; very often, debugging becomes a pain. Developers had to partially debug in a browser, and native features debug with `console.log` on devices. Debugging is not so convenient as in Chrome, Safari, or Firefox.

Here is where GapDebug from Genuitec comes in very handy. GapDebug is a free cross-platform mobile app debugger for use with Android and iOS PhoneGap (or Cordova) applications. You can download and install it from `https://www.genuitec.com/products/gapdebug/download/`. It is available for both Windows and Mac OS. The installation itself is very clear and straightforward. So, I don't think there is a need to explain it here. We will explain how it makes debugging hybrid applications deployed with PhoneGap easier than ever before.

Here are some mobile OS and configuration requirements:

	Android	**iOS**
Mobile device	<ul><li>Android 4.4+ (KitKat or greater) device or emulator</li><li>USB debugging enabled</li><li>Enabled debugging</li></ul>	<ul><li>iOS 7 or 6</li><li>Web Inspector enabled</li><li>App must be signed with an iOS Developer certificate</li></ul>
Computer: Windows 7 or 8, and Mac OSX	<ul><li>Android USB driver (Windows only)</li><li>Chrome browser</li></ul>	<ul><li>iTunes installed</li><li>Chrome browser</li></ul>

iOS debugging setup

We need to follow some steps before we can start debugging with GapDebug.

Computer configuration

We need to prepare a development computer. To do this:

- Install Chrome browser if it is not already installed — GapDebug uses its interface

- Install iTunes — GapDebug uses its USB driver

iOS device configuration

We need to enable Web Inspector in Safari on an iOS device. To do this, follow these steps:

1. Open **Settings** and click on the **Safari** menu item.
2. In the **Safari** section, click on **Advanced**.
3. Enable **Web Inspector**.

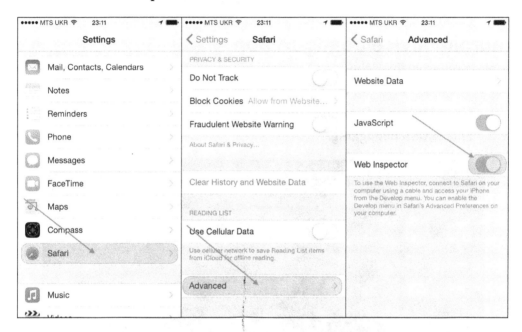

 USB debugging is available only for iOS 6 and 7 applications that have been signed with a Developer certificate.

Android debugging setup

We need to follow some steps before we can start debugging with GapDebug.

Computer configuration

We need to prepare a development computer. To do this:

- Install Chrome browser if it is not already installed. GapDebug uses its interface.

- Install the Android USB driver (Windows only). The Windows version of GapDebug requires the Android USB driver to communicate with connected Android devices. There is no need to install such a driver for Mac OS X.

Android device configuration

We also need to enable debugging through USB on an Android device. To do this, follow these steps:

1. Go to **Settings | Developer options**.
2. Enable **USB debugging**.

In some OSes, we need to unlock developer options first. For example, on Android Jelly Bean, we need to tap the version number five times to unlock it.

When we fulfill all the requirements, we simply need to plug and debug our device by connecting it to the GapDebug host computer using a USB cable. Alternatively, we can debug applications running in the emulator. I have only an iPhone; I do not have an Android device. I debug an iOS application version on a real device and an Android application on the emulator. Once the connection is completed, the device information will appear to the left of the GapDebug user interface in the Device Panel.

Here is how the initial screen of the GapDebug looks for me with an Android emulator and a connected iPhone:

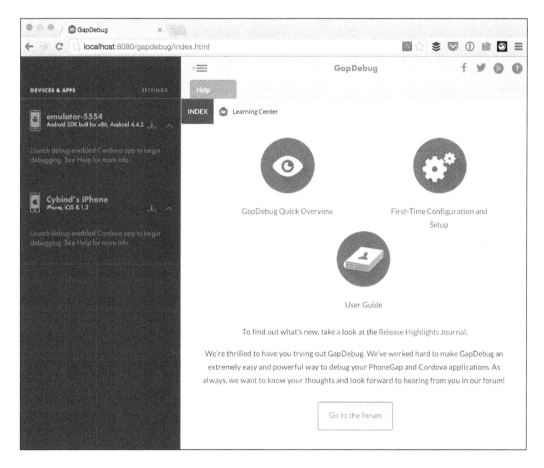

On the left-hand side, you can see two active devices: **emulator-5554**, which we got configured in the previous chapter, and **Cybind's iPhone**, which I have just attached to my computer.

Now, we should be able to run our PhoneGap application on both the device and emulator.

The official GapDebug documentation says: *On the mobile device, launch the app that you wish to debug. If the app is not already present or up to date on the device, use the app installation button found next to the device info to select and install a new version of the app. Alternatively, you can drag and drop a binary app executable file on the device to install it.*

In our case, since we are using a Sencha Touch Cmd, it is enough to go to the working directory and execute the following command:

```
$ sencha app build -run native
```

When the app is launched, the app's ID and icon will appear just under the device in the GapDebug Device Panel:

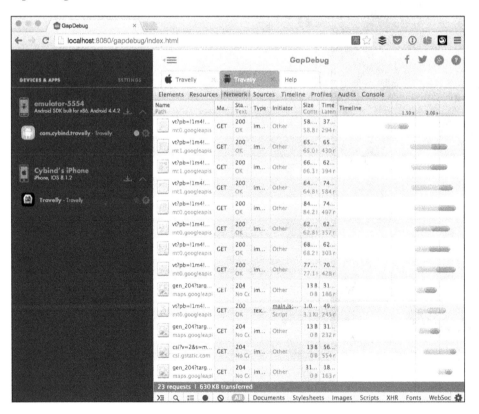

Using GapDebug, you can simultaneously debug multiple apps running on iOS and Android.

GapDebug has HTML and CSS inspection, JavaScript debugging, and resource profiling capabilities. That is why I think it is the best tool to debug PhoneGap/Cordova applications in current time.

The Genymotion Android emulator for faster debugging

When you develop applications for Android and test your application in an Android emulator, you might realize that it is painfully slow, especially when you look at animations or other complex effects. Even navigation is not responsive. Of course, you can install the HAX module for your emulator, but it does not add much performance as well.

You can read more about HAX installation and configuration at `https://software.intel.com/en-us/android/articles/installing-the-intel-atom-x86-system-image-for-android-emulator-add-on-from-the-android-sdk-manager`.

One day, I found a mention of the Genymotion Android emulator.

> *Genymotion is a fast and easy-to-use Android emulator to run and test your Android apps.*

It uses CPU and OpenGL acceleration; that is why it is very fast. I have installed it, and it is really great. I recommend you to try it. I believe you will like it and will use it for a long time.

You can download it from `https://www.genymotion.com/`. The official website has very good documentation on its installation and configuration.

The initial application's MVC structure

After preparing some debugging tools, we can move back to our "Travelly" application development itself.

We should organize our views, controllers, and models. Let's start with views.

Views

Earlier, when we created a basic Sencha Touch application, we described the launch method in the main entry point:

```
launch: function() {
    Ext.fly('appLoadingIndicator').destroy();
    Ext.Viewport.add(Ext.create('Travelly.view.Main'));
}
```

This code destroys the loading indicator and adds our `Main` view to the viewport. A loading indicator displays the process when the application is not ready yet and initiating all the components. This is pretty clear, right? By default, the `Main` view extends the `Ext.tab.Panel` Sencha component.

Tab panels are a great way to allow the user to switch between several pages that are all full screen. Each component in the tab panel gets its own tab, which shows the component when tapped on it. Tabs can be positioned at the top or the bottom of the tab panel and can optionally accept title and icon configurations.

And it works well for our needs.

In the header of the view, let's specify the `main` settings:

```
xtype: 'main',
requires: [
    'Ext.TitleBar',
    'Ext.Button',
    'Ext.Img',
    'Ext.Map'
]
```

The preceding code consists of the following elements:

- `xtype`: This is an alias to reference in other places; for example, in other views
- `Ext.TitleBar`: This is the bar with the title that we will use on every tab panel
- `Ext.Button`: This is a simple class to display a button
- `Ext.Img`: This is an easy way to add an image to the application
- `Ext.Map`: This is a Google map for our future needs

After the `requires` sections, we will add only one other section, which is `config`:

```
config: {
    tabBarPosition: 'bottom',
    items: []
}
```

Here, each property has the following meaning:

- `tabBarPosition`: This is an option where we specify the docked position of the tabBar instance. In our case, we place it at the `bottom`
- `items`: This is where we will add the child elements to the container

Let's add two tabs: **New Photo** and **My Places**. Every tab panel will have a title bar. Let's take a close look at the first tab panel:

```
{
    title: 'New Photo',
    iconCls: 'lens',
    items: [
        {
            docked: 'top',
            xtype: 'titlebar',
            title: 'New Photo'
        },
        {
            xtype: 'container',
            width: '100%',
            height: '100%',
            layout: {
                type: 'vbox',
                pack: 'center',
                align: 'center'
            },
            items: [{
                    xtype: 'button',
                    id: 'takePhotoBtn',
                    text: 'Take Photo',
                    iconCls: 'photo',
                    iconAlign: 'top',
                    height: 65,
                    width: 120,
                    padding: 10,
                    margin: 5
            }]
```

```
            }
        ]
    }
```

We just created a tab panel with the New Photo button and the lens icon at the bottom. The panel contains two items: `titlebar` and `container`.

The `titlebar` item is docked at the top and has the text New Photo inside. The `container` item itself fills the entire space and has a vbox layout. The vbox layout is one of the most useful layouts in Sencha Touch, as it can arrange components in a wide variety of vertical combinations. The `pack` and `align` features control how child elements are aligned in a layout. The `pack` feature refers to the axis of our current layout, while `align` is the opposite. In our case, in a vbox layout, `pack` refers to the vertical axis and `align` refers to the horizontal axis.

In the `items` array, we have only one item: `button`. We want to open the camera once the user clicks this button. We have assigned an ID to the button so that we can refer to it in our controller and handle events. All other options of the button are UI parameters. The `text` property specifies the title on the button, `iconCls` is a CSS class to add to the icon element, and `iconAlign` is a position of the icon in the button to be rendered. The `height`, `width`, `margin`, and `padding` properties are well known in CSS developer styles, and it is clear what they represent.

On the second tab, we will display Google map. On the map, we will show markers of the places where we took a photo with our camera. Let's take a close look at the second tab panel:

```
{
    title: 'My Places',
    iconCls: 'globe',
    items: [
        {
            docked: 'top',
            xtype: 'titlebar',
            title: 'Places I visited'
        },
        {
            xtype: 'map',
            id: 'mapPlaces',
            useCurrentLocation: true,
            width: '100%',
            height: '100%'
        }
    ]
}
```

For this tab panel, we assigned a button with the title `My Places` and with the icon `globe`. The `titlebar` item displays the `Places I visited` text. Inside the panel, we have placed full `width` and `height` for Google map. The `useCurrentLocation` property tells the map to scroll to the location where the device currently exists.

Adding Pictos icons to the application

You might realize that for some components, we used the `iconCls` parameter to assign icons with CSS. By default, Sencha Touch uses the Pictos font for icons.

Pictos offers hand-crafted, infinitely scalable, and royalty-free icons for user interface designers.

Icons are handled as fonts, which enables very fast scaling without the need to redraw the picture. Icons appear on buttons and tabs. When we build our application using Sencha Cmd, the font libraries are copied to our application file. Sencha Touch conveniently maps icon names to actual icons so that we can use them within our application. You can find a full list of characters available for the Pictos font at `http://pictos.cc/classic/font`:

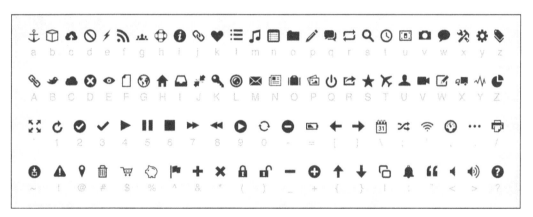

We used the compass icon mixin to map a specific character of the icon font to `iconCls`. We added the following lines of code to `resources/sass/cupertino.scss`:

```
@include icon('lens', 'L');
@include icon('photo', 'v');
@include icon('globe', 'G');
@include icon('check', '3');
```

After that, just run the following command:

```
$ sencha app watch
```

 SCSS is a syntax for SASS. SASS itself is a CSS extension language. You can read more about it at `http://sass-lang.com/`.

It will compile CSS for us. After that, we can use a lens, photo, and globe CSS classes to assign icons to our buttons.

Eventually, we will get an application that will look similar to this screenshot:

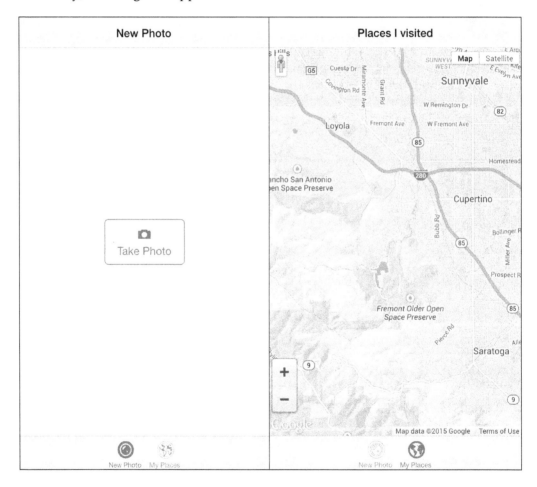

For now, the UI is the same for iOS and Android. When we click between tabs on the bottom panel, the app view switches between the **New Photo** and **Places I visited** screens. The **Take Photo** button does nothing for now.

Controllers

In the initial stage, we have a single controller, `Travelly.controller.Main`. Let's add an empty handler for our **Take Photo** button on the main application screen:

```
Ext.define('Travelly.controller.Main', {
    extend: 'Ext.app.Controller',
    config: {
        refs: {
            takePhotoBtn: '#takePhotoBtn',
            mapPlaces: '#mapPlaces'
        },
        control: {
            takePhotoBtn: {
                tap: 'getPhoto'
            }
        }
    },
    getPhoto: function() {
        // get photo code goes here...
    }
});
```

We just created references to the **Take Photo** button and map component on the **Places I visited** tab panel, in the same way we described in the previous chapter. We simply attached an event handler `getPhoto` on the tap event for the **Take Photo** button.

Model and store

In the previous chapter, we already described the picture model and store. In this case, we will only remove inline data from our store so that it looks like this:

```
Ext.define('Travelly.store.Pictures',{
    extend:'Ext.data.Store',
    requires: ['Ext.data.proxy.LocalStorage',
    'Travelly.model.Picture'],
    config:{
        model:'Travelly.model.Picture',
```

```
        storeId: 'Pictures',
        autoLoad: true,
        autoSync: true,
        proxy:{
            type:'localstorage'
        }
    }
  });
```

We do this to allow data modification in the store so that we do not have inline data and the same data all the time.

So, it is time now to learn something about PhoneGap/Cordova plugins and their interaction with devices.

Using the Cordova StatusBar plugin to fix overlap

When you start the application for the first time on an iOS device, you will realize that there is some issue with the status bar when the application is running. Here is how the status bar looks on my device:

You can see that the text in the status bar is white, and only the battery indicator is green. Also, you can see that the status bar overlaps our title bar. This is due to the iOS 7 change for ViewControllers (containing `WebView` running our PhoneGap application) to display full screen by default with a transparent status bar overlaying them rather than having their own designated space in the top 20 pixels of the screen as you saw earlier. There are two possible solutions:

- Provide CSS adjustments to the title bar; add some margin at the top and make its background dark
- Use the Cordova status bar plugin to do it natively

I prefer to use the second approach, because the first one looks more like cheating than a good solution.

To install the Cordova plugin, we have to find the proper plugin first. We can do this by changing our current folder to `cordova` and running the `plugin search` command:

```
$ cd cordova/
$ cordova plugin search statusbar
```

It is looking for an appropriate plugin in the Apache Cordova plugin registry at `http://plugins.cordova.io/`. It returns a lot of results, but we need to install only `org.apache.cordova.statusbar`. To do this, let's run the following command:

```
$ cordova plugin add org.apache.cordova.statusbar
```

This command downloads and installs the plugin with NPM. The CLI adds the plugin code as appropriate for each platform. Now, if you go to `cordova/plugins/org.apache.cordova.statusbar/src`, you will see that there is native code for different platforms:

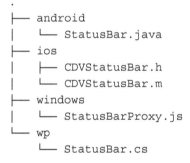

```
.
├── android
│   └── StatusBar.java
├── ios
│   ├── CDVStatusBar.h
│   └── CDVStatusBar.m
├── windows
│   └── StatusBarProxy.js
└── wp
    └── StatusBar.cs
```

In the www folder, you will see a single `statusbar.js` file, which is included in the HTML part of the application. From this folder, files are copied to the appropriate platform if available. For example, for the iOS platform, `cordova/plugins/org.apache.cordova.statusbar/src/ios` is copied to the following location:

```
cordova/platforms/ios/Travelly/Plugins/org.apache.cordova.statusbar
```

And `cordova/plugins/org.apache.cordova.statusbar/www/statusbar.js` is copied to:

```
cordova/platforms/ios/www/plugins/org.apache.cordova.statusbar/www/
statusbar.js
```

Also, to include JavaScript files of the plugin in the application, we used `cordova/platforms/ios/www/cordova_plugins.js`. All installed plugins are listed in the following format:

```
{
    "file": "plugins/org.apache.cordova.statusbar/www/statusbar.js",
    "id": "org.apache.cordova.statusbar.statusbar",
    "clobbers": [
        "window.StatusBar"
    ]
}
```

Here, `clobbers` defines an array of references where this plugin will be available in the JavaScript part of our application. For example, the `statusbar` plugin will be available under `window.StatusBar`.

The `StatusBar` plugin can be used to natively address the overlap issue by setting the `StatusBarOverlaysWebView` preference to `false` in `config.xml`. By default, this preference is set to `true`. That is why you will not be able to see any changes to the application right after plugin installation. Setting `StatusBarOverlaysWebView` to `false` also allows a color to be applied to the background. Here is what we add to `cordova/config.xml`:

```
<preference name="StatusBarOverlaysWebView" value="false" />
<preference name="StatusBarBackgroundColor"
value="#000000" />
<preference name="StatusBarStyle" value="lightcontent" />
```

Where:

- `StatusBarBackgroundColor` sets the background color of the bar only when `StatusBarOverlaysWebView` is `false`

- `StatusBarStyle` sets the foreground color, including the text and icons

Status bar styles that are available include the following ones:

- The `default` and `blacktranslucent` values display black foreground text and icons

- The `lightcontent` and `blackopaque` values display white foreground text and icons

As a result, we should see something like this:

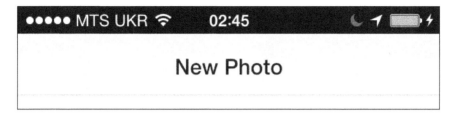

Once we have fixed the iOS status bar, let's go to the camera plugin and integrate it with our application.

Using a camera to capture pictures

Now, we would like to take pictures with a camera inside our application, attach a caption to it, and save it for future usage. First of all, let's install the Cordova plugin to access the camera.

Camera plugin installation

To install the plugin, simply run this command:

```
$ cordova plugin add org.apache.cordova.camera
```

The execution of the command will take a while. At the end of the flow, you will see the following messages:

```
Installing "org.apache.cordova.camera" for android
Installing "org.apache.cordova.camera" for ios
```

Now, we have this plugin installed for both iOS and Android platforms.

> The Cordova Camera plugin supports Amazon Fire OS, Android, BlackBerry 10 browser, Firefox OS, iOS, Tizen, Windows Phone 7 and 8, and Windows 8 platforms.

This plugin defines a global navigator, a camera object, that provides an API to take pictures and choose images from the system's image library. Let's look at it using an example.

Camera plugin usage

Let's go to our `Travelly.controller.Main` controller placed under `app/controller/Main.js` and add this method:

```
getCameraPicture: function(callback) {
    if (Ext.browser.is.PhoneGap) {
        var onSuccess = function(imageURI) {
            if (callback) callback(imageURI);
        }
        var onFail = function(message) {
            alert('Failed because: ' + message);
            if (callback) callback();
        }
        navigator.camera.getPicture(onSuccess, onFail, {
            quality: 50,
            destinationType:
            navigator.camera.DestinationType.FILE_URI,
            sourceType: navigator.camera.PictureSourceType.CAMERA,
            encodingType: navigator.camera.EncodingType.JPEG,
            correctOrientation: true
        });
    } else {
        // Emulate captured image
    }
}
```

As a parameter, we passed `callback` to the function. The `callback` parameter represents a function that we will call once the camera's async execution function is finished. With `Ext.browser.is.PhoneGap`, we checked whether we are running the application on a real device/emulator or in a browser. If we are not able to identify whether it is running the application, it means we started our application in a browser and our camera device's function will not work. For debugging needs in a browser, we can mockup a picture instead of taking it from a camera. I develop the entire HTML layout mainly in a browser, and only when I need to test native device functions, I run it in emulators. I would recommend you to do it the same way.

Once we checked that the application is running on a real device/emulator, we called the `navigator.camera.getPicture` function immediately. The function opens the device's default camera application that allows users to snap pictures or select a photo from the device's image gallery. The image is passed to the success callback as a base64-encoded string or as the URI for the image file. This function has several parameters, which are as follows:

- `quality`: This is the quality of the saved image. The default is `50`, but it can be from `0` to `100`, where `100` is full resolution.

- `destinationType`: This defines the format of the return value. By default, it is `DestinationType.FILE_URI`, but other possible options are `DestinationType.FILE_URI` and `DestinationType.NATIVE_URI`.

- `DATA_URL`: This returns the image as a Base64-encoded string.

  Base64 is a group of similar binary-to-text encoding schemes that represent binary data in an ASCII string format by translating it into a radix-64 representation. The term Base64 originates from a specific MIME content transfer encoding.

- `FILE_URI`: This returns the image file URI.

- `NATIVE_URI`: This returns the image's native URI.

- `sourceType`: This allows us to change the location from which we should take a picture: using the camera or from the device's library. By default, it is `PictureSourceType.CAMERA`, but it can be `PictureSourceType.PHOTOLIBRARY` or `PictureSourceType.SAVEDPHOTOALBUM`.

- `encodingType`: This allows us to select the returned image file's encoding. By default, it is `EncodingType.JPEG`, but the other option is `EncodingType.PNG`.

- `correctOrientation` rotates the image to correct the orientation of the device during capture.

 Photo resolution on newer devices is quite good. Photos selected from the device gallery are not downscaled to a lower quality, even if a quality parameter is specified. To avoid common memory problems, set `destinationType` to `FILE_URI` rather than `DATA_URL`.

The first two parameters of the `navigator.camera.getPicture` function are `onSuccess` and `onFail` callbacks. If we get a success callback, that is great. We pass the returned parameter (in our case, it is the image URI) further in the application. However, if we get an error and forward it to the `onFail` callback, we can check what the error message is.

There are several other camera options we have not used in our application, but you can look at them on the official plugin page at `http://plugins.cordova.io/#/ package/org.apache.cordova.camera`.

Once the user snaps the photo, the camera application closes, and the application is restored.

Now, let's link our `getCameraPicture` function with a tap event on the **Take Photo** button. Earlier in the chapter, we already defined an empty `getPhoto` method and attached it as a handler on the tap event for the button. Now, we will modify it to look something like this:

```
getPhoto: function() {
    var self = this;
    self.getCameraPicture(function(imageURI) {
        self.showPhotoPopup(imageURI);
    });
}
```

As you can see, we passed an anonymous function as a parameter of the `getCameraPicture` function. This function receives an `imageURI` parameter pointing to the picture we created with the camera. Next, we will use the `showPhotoPopup` function to display a full-screen popup with picture preview, a text field to enter the picture caption, and three buttons: **Retake**, **Save**, and **Cancel**. Let's define the view for the popup first.

Creating a new picture popup

Let's create the `app/views/NewPicture.js` view with the following content:

```
Ext.define('Travelly.view.NewPicture', {
    extend: 'Ext.Panel',
    xtype: 'newpicture',
    requires: [
        'Ext.TitleBar',
        'Ext.Button',
        'Ext.Img',
        'Ext.field.Text'
    ],
```

```
config: {
    height: '100%',
    width: '100%',
    centered: true,
    showAnimation: 'slideIn',
    hideAnimation: 'slideOut',
    hidden: true,
    items: []
}
});
```

It is the standard Ext.Panel Sencha Touch panel with some configuration. We added its xtype property for easy reference in the future. We will use the title bar, buttons, image, and text field components on the page. That is why we added all the required components in the required sections.

In the config section, you can see that we specified a full-width and full-height centered panel. The showAnimation and hideAnimation animation effects can be applied when the panel is being shown.

On clicking on the show event, the panel will slide in from the right-hand side, and on clicking on the hide event, it will slide out to the left-hand side. By default, we make it hidden. So, when we add it to the view port, it is not displayed immediately. It slides in only when we click on show event.

The next step is to add child components to our panel. We want to create a combination of horizontal and vertical layouts, so will use additional containers to align buttons, pictures, and text fields as desired:

```
{
    docked: 'top',
    xtype: 'titlebar',
    title: 'New Photo'
},
{
    xtype: 'container',
    width: '100%',
    height: '100%',
    layout: {
        type: 'vbox',
        pack: 'center',
        align: 'center'
    },
    items: []
}
```

Here, we added our title bar and full-size vbox container. Everything in the container will be horizontally and vertically centered. In the container, we will add image and textfield components for horizontally aligned buttons:

```
{
    xtype: 'image',
    id: 'photoPreview',
    width: '100%',
    flex: 8
},
{
    xtype: 'textfield',
    id: 'photoTitle',
    clearIcon: true,
    width: '90%',
    margin: '5 0 0 0',
    flex: 1
},
{
    xtype: 'container',
    flex: 1,
    layout: {
        type: 'hbox',
        pack: 'center',
        align: 'center'
    },
    items: []
}
```

You might notice a new parameter called flex in the components. Flexing means that we divide the available area based on the flex of each child component. These three components are vertically aligned, so the image will take up 80 percent of the height, the text field will be 10 percent, and the container for buttons will be 10 percent. The image and textfield components have IDs, so we can refer to them in the preceding code.

- xtype: 'image': This is a simple way to add an image of any size to our application and have it participate in the layout system like any other component. This component typically takes between one and three configurations: a src and, optionally, a height and a width.

- xtype: 'textfield': The textfield is the basis for most of the input fields in Sencha Touch. It provides a baseline of shared functionality such as input validation, standard events, state management, and look and feel.

Now, let's add three buttons to the button's container, as follows:

```
{
    xtype: 'button',
    id: 'retakePhotoBtn',
    text: 'Retake',
    iconCls: 'photo',
    flext: 1,
    margin: '0 5 0 5'
},
{

    xtype: 'button',
    id: 'savePhotoBtn',
    text: 'Save',
    iconCls: 'check',
    flext: 1,
    margin: '0 5 0 5'
},
{

    xtype: 'button',
    id: 'cancelPhotoBtn',
    text: 'Cancel',
    iconCls: 'delete',
    flext: 1,
    margin: '0 5 0 5'
}
```

These are three simple same-width buttons with a 5 pixel margin on the left and right sides. Tapping these buttons triggers the following actions:

- `retakePhotoBtn tap`: This opens the native camera application to capture a picture one more time
- `savePhotoBtn tap`: This saves the picture and other related information in the application and closes the popup
- `cancelPhotoBtn tap`: This simply closes the popup

Let's take care of implementing these actions in the main controller. Once we click on the **Take Photo** button, it opens the native camera application where we take a picture. In the `getCameraPicture` callback, we will call the `showPhotoPopup` function where we pass the `imageURI` parameter that we took:

```
showPhotoPopup: function(imageURI) {
    var self = this;
```

```
var popup = Ext.create('Travelly.view.NewPicture');
Ext.Viewport.add(popup);
popup.show();
popup.on('hide', function() {
    popup.destroy();
});
self.setPreviewImage(imageURI);
Ext.getCmp('retakePhotoBtn').on('tap', function() {
    self.getCameraPicture(self.setPreviewImage);
});
Ext.getCmp('savePhotoBtn').on('tap', function() {
    var title = Ext.getCmp('photoTitle').getValue();
    self.savePhoto(imageURI, title);
    popup.hide();
});
Ext.getCmp('cancelPhotoBtn').on('tap', function() {
    popup.hide();
});
}
```

In the function, we created an instance of the `Travelly.view.NewPicture` view, added it to the view port, and triggered the show event. On the hide event, we destroyed the popup. We did this to avoid popup duplicates and memory leaks related to it. Thus, the next time we open the popup, the previous one is already destroyed. The `Ext.getCmp` function looks up a button by ID, and the `on` function assigns event handler for `tap`.

When the user clicks on the `retakePhotoBtn` button, we call `self.getCameraPicture` again, but as a callback passing another function:

```
setPreviewImage: function(imageURI) {
    Ext.getCmp('photoPreview').setSrc(imageURI);
}
```

This function simply finds the image component with the `photoPreview` ID and assigns a new `imageURI` parameter to it. It means that, once we get a new picture from the device's camera, it will be placed instead of the previous image in the popup.

When the user clicks on the `savePhotoBtn` button, we take text from the text field and pass it along with `imageURI` to the `savePhoto` function. We will describe this function a little later.

By clicking on the `cancelPhotoBtn` button, we simply call the `popup.hide()` method to hide the popup.

We should remember that, to use created views, we should add them to the `views` section in `app.js` so that it looks like this:

```
views: [ 'Main', 'NewPicture' ]
```

By that moment of application development, the popup should look like this screenshot:

Now let's take a deeper look at the `savePhoto` function.

Filesystem plugin installation and usage

The problem with a native camera is that it stores captured pictures in a temporary folder. For example, the `imageURI` from the camera plugin taken on an iPhone might look like this: `file:///var/mobile/Containers/Data/Application/AB784267-B994-4B54-A81C-EF1180B18738/tmp/cdv_photo_005.jpg`.

Next time you run the application, this `tmp` folder might be cleared. Here are a few possible solutions to the issue:

- Use base64 data of the picture and store in the internal application's database
- Move the picture to the persistent store where mobile OS is not able to remove files if more space is needed

The first option doesn't work because there is a limit for database file size and, on modern devices, pictures can be really high-quality and, thus, large. I tried to use this approach and was able to save only three pictures on an iPhone 5. Four pictures were taking more space in the database than is allowed.

So, I followed the second approach. There is a good Cordova plugin to work with the device's filesystem: `org.apache.cordova.file`. This plugin implements a File API, allowing read/write access to files residing on the device.

 You can read more about the File API on W3C at `http://dev.w3.org/2009/dap/file-system/pub/FileSystem/`. The Cordova File plugin supports Amazon Fire OS, Android, BlackBerry 10, Firefox OS, iOS, Tizen, Windows Phone 7 and 8, and Windows 8 platforms.

Let's install the plugin using this command:

```
$ cordova plugin add org.apache.cordova.file
```

Now, we can use this plugin to copy our pictures into a permanent folder.

Using a persistent file location

Currently, our `savePhoto` function looks like the following code:

```
savePhoto: function(imageURI, title) {
    var self = this;
    self.copyPhotoToPersistentStore(imageURI,
    function(persistentImageURI) {
```

```
        // get current location and save goes here
    })
}
```

It is nothing complicated; only `copyPhotoToPersistentStore` has its own implementation with the File API usage:

```
copyPhotoToPersistentStore: function(fileURI, callback) {
    var onSuccess = function(entry) {
        var d = new Date();
        var n = d.getTime();
        var newFileName = n + ".jpg";
        var myFolderApp = "Travelly";
        window.requestFileSystem(LocalFileSystem.PERSISTENT, 0,
        function(fileSys) {
            fileSys.root.getDirectory(myFolderApp, {
                    create: true,
                    exclusive: false
                },
                function(directory) {
                    entry.copyTo(directory, newFileName,
                    successCopy, onError);
                },
                onError);
        },
        onError);
    }

    var successCopy = function(entry) {
        if (callback) callback(entry.nativeURL);
    }

    var onError = function(error) {
        alert(error.code);
        if (callback) callback();
    }

    window.resolveLocalFileSystemURL(fileURI, onSuccess, onError);
}
```

The main entry point where `imageURI` passed is `window.resolveLocalFileSystemURL`. This function gets back the file entry using the file URI and passes it to the `onSuccess` callback. There, we prepared a new picture name as a JavaScript Date value and defined the folder name as `Travelly`, where we will place the picture. A Web app can request access to a device filesystem by calling `window.requestFileSystem(type, size, successCallback, opt_errorCallback)`, where:

- `type`: This defines whether the file storage should be persistent or not. The possible values are `LocalFileSystem.TEMPORARY` or `LocalFileSystem.PERSISTENT`. Data stored using `TEMPORARY` can be removed at the device's discretion (for example, if more space is needed).

- `size`: This is the size (in bytes) the app will require for storage. `0` means not defined.

- `successCallback`: This is a callback that is invoked on the successful request of a filesystem. Its argument is a `FileSystem` object.

- `opt_errorCallback`: This is an optional callback to handle errors or when the request to obtain the filesystem is denied. When an error is thrown, one of the following codes will be used:

  - `NOT_FOUND_ERR`
  - `SECURITY_ERR`
  - `ABORT_ERR`
  - `NOT_READABLE_ERR`
  - `ENCODING_ERR`
  - `NO_MODIFICATION_ALLOWED_ERR`
  - `INVALID_STATE_ERR`
  - `SYNTAX_ERR`
  - `INVALID_MODIFICATION_ERR`
  - `QUOTA_EXCEEDED_ERR`
  - `TYPE_MISMATCH_ERR`
  - `PATH_EXISTS_ERR`

So, we defined that we need to get access to the persistent store. We did not specify the required space and assigned success and failure callbacks. In the success callback, we receive the `FileSystem` object into the `fileSys` variable. The `fileSys.root.getDirectory` function creates a `Travelly` directory in the application's root directory and returns the directory entry in the success callback. Now, we can copy our file entry into a new directory using the directory entry. `FileEntry` and `DirectoryEntry` share common operations. Both `FileEntry` and `DirectoryEntry` have a `copyTo()` function to duplicate existing entries. It copies the file for us and returns a new file entry in the success callback. The file entry has several properties, but most important for us is the native URL. If you look into its value, you will see something like this: `file:///var/mobile/Containers/Data/Application/AB784267-B994-4B54-A81C-EF1180B18738/Documents/Travelly/1420758818629.jpg`.

Now, the file is stored in the `Documents` persistent folder. It is exactly the URI we return with our callback from the `copyPhotoToPersistentStore` function.

> You can see the content of the variables using the GapDebug tool. Simply use `console.log(entry.nativeURL)` or place a break point to track the variable.

The next step is to detect our current location and save all the collected information.

Detecting the current geolocation

It is very easy to detect a device's current location (latitude and longitude) with the Cordova `org.apache.cordova.geolocation` plugin.

> Details of the Geolocation API are available on the W3C site at `http://dev.w3.org/geo/api/spec-source.html`.

This plugin detects the current location based on GPS, IP address, RFID, Wi-Fi and Bluetooth MAC addresses, and GSM/CDMA cell IDs. However, there is no guarantee that the location will be detected precisely enough.

Let's install it the same way we did for other plugins:

```
$ cordova plugin add org.apache.cordova.geolocation
```

Now, let's modify our `savePhoto` function in the following way:

```
savePhoto: function(imageURI, title) {
    var self = this;
    self.copyPhotoToPersistentStore(imageURI,
    function(persistentImageURI) {
        self.getCurrentPosition(function(latitude, longitude) {
            // saving goes here
        })
    })
}
```

As you can see, we added another function called `self.getCurrentPosition` here. It receives latitude and longitude as parameters in the callback. This is how the function looks:

```
getCurrentPosition: function(callback) {
    var onSuccess = function(position) {
        callback(position.coords.latitude,
        position.coords.longitude);
    }
    var onError = function(error) {
        alert('code: '    + error.code    + '\n' +
              'message: ' + error.message + '\n');
    }
    navigator.geolocation.getCurrentPosition(onSuccess, onError);
}
```

A single function in the method is `navigator.geolocation.getCurrentPosition`. The `onSuccess` callback accepts a `position` object, which contains the current GPS coordinates. In addition, it has altitude accuracy, heading, speed, and timestamp.

The `onError` failure callback receives these possible errors:

- `PERMISSION_DENIED`: This is returned when users do not allow the application to retrieve position information.

- `POSITION_UNAVAILABLE`: This is returned when the device is unable to retrieve a position. In general, this means the device is not connected to a network or can't get a satellite fix.

- `TIMEOUT`: This is returned when the device is unable to retrieve a position within the time specified by the timeout included in `geolocationOptions`.

Saving data in local storage

We have captured a picture, assigned a caption, moved the picture to the persistent store, and detected the picture's location. Now, we need to save this information into the application's database:

```
savePhoto: function(imageURI, title) {
    var self = this;
    self.copyPhotoToPersistentStore(imageURI,
    function(persistentImageURI) {
        self.getCurrentPosition(function(latitude, longitude) {
            var picture = Ext.create('Travelly.model.Picture', {
                url: persistentImageURI,
                title: title,
                lat: latitude,
                lon: longitude
            });
            var pictureStore = Ext.getStore('Pictures');
            pictureStore.add(picture);
            pictureStore.sync();

            var map = self.getMapPlaces().getMap();
            Travelly.app.getController('Places')
            .addMarker(picture, map);
        });
    })
}
```

Here, we created the `Picture` model with the persistent image URI, title, and coordinates. We then added it to the `Pictures` store and synced it. The next two rows of code relate to the map on the second tab panel in the application. We are calling the `addMarker` function of the Places controller, which adds the marker to the Google map. The marker represents a place on the map where we took the shot. Let's look at this controller in detail.

Displaying data with Google Maps

Now, it is time to implement the functionality to display pictures we created with our application. It would be nice to display markers on the map for the locations where we created the picture. Let's create a new controller called `Places`:

```
$ sencha generate controller Places
```

This command creates the `app/controllers/Places.js` file for us. Let's modify it a little bit to look like this:

```
Ext.define('Travelly.controller.Places', {
    extend: 'Ext.app.Controller',
        config: {
        refs: {
            mapPlaces: '#mapPlaces'
        },
        control: {
            mapPlaces: {
                maprender: 'mapPlacesMapRenderer'
            }
        }
    },
    mapPlacesMapRenderer: function(comp, map) {
        var pictures = this.getPictures();
        var map = this.getMapPlaces().getMap();
        for (var i = 0; i < pictures.length; i++) {
            this.addMarker(pictures[i], map);
        }
    },
    ...
});
```

It is a standard Sencha Touch controller with reference to the `map` component on the **Send** tab in our application. The map fires the `maprender` event when the map is initially rendered. We use this event to handle the state when the map is ready and we can add markers to it. In the `mapPlacesMapRenderer` function, first of all we retrieve pictures from the database with the following function in the same `Places` controller:

```
getPictures: function() {
    var pictureStore = Ext.getStore('Pictures');
    var data = pictureStore.getData();
    return data ? data.all : null;
}
```

Here, we got an instance of our pictures store by an alias name that retrieves all the data with the getData() function. Stores loads data via a Proxy, and also provides functions to sort, filter, and query the model instances contained within it. However, in our case, it is enough to retrieve it in the default order, loop through the models array, and put to the map with:

```
addMarker: function(picture, map) {
    var self = this;
    var marker = new google.maps.Marker({
        map        : map,
        position   : new google.maps.LatLng(picture.get('lat'),
        picture.get('lon')),
        title      : picture.get('title'),
        icon       : 'resources/images/camera.png',
        picture    : picture
    });
}
```

Here, picture is the Picture model from the database, and map is an instance of our map component. When we call the Google map function to create a marker, we are passing several parameters:

- map (optional): This specifies the map on which to place the marker. If we do not specify the map on construction of the marker, the marker is created but is not attached to (or displayed on) the map.

- position (required): This specifies a LatLng property identifying the initial location of the marker.

- title: This is a marker's title. It will appear as a tooltip. We assign it as the picture's caption we entered when we created it.

- icon: This indicates an image to use instead of the default Google Maps pushpin icon.

- picture: This is our custom property where we attach our property object for later reference to it when we open picture details.

Once we get it implemented, we should not forget to add our Places controller to the app.js file:

```
controllers: [ 'Main', 'Places' ]
```

Once everything is ready, we will be able to see markers on the map:

You can see a small marker with a camera icon on it. It is the location where I took a shot and saved the picture.

Next, we will preview the picture. It will open once the user clicks on the marker.

Displaying picture details in a popup

We want to display the picture and its title on a full-screen popup, along with the option to remove the picture from the filesystem and related information from the database.

We will create a view similar to the full-screen popup we created to save new pictures. Let's create this view in app/views/Picture.js.

The main components in the view are:

- Title bar to display the saved picture caption:

```
{
    docked: 'top',
    xtype: 'titlebar',
    id: 'photoTitle',
    title: ''
}
```

- Image to display the picture taken at the selected place:

```
{
    xtype: 'image',
    id: 'photoPreview',
    width: '100%',
    flex: 9
}
```

- Two buttons to delete the picture and close the popup:

```
{
    xtype: 'button',
    id: 'deletePhotoBtn',
    text: 'Delete',
    iconCls: 'trash',
    ui: 'decline',
    flext: 1,
    margin: '0 5 0 5'
},
{
    xtype: 'button',
    id: 'closePhotoBtn',
    text: 'Close',
    iconCls: 'delete',
    flext: 1,
    margin: '0 5 0 5'
}
```

To be able to open this popup, let's add a `click` event handler to each marker:

```
addMarker: function(picture, map) {
    ...
    google.maps.event.addListener(marker, 'click', function() {
        var marker = this;
```

```
var popup = Ext.create('Travelly.view.Picture');
Ext.Viewport.add(popup);
popup.show();

popup.on('hide', function() {
    popup.destroy();
});

Ext.getCmp('photoPreview')
.setSrc(marker.picture.get('url'));
Ext.getCmp('photoTitle')
.setTitle(marker.picture.get('title'));

Ext.getCmp('deletePhotoBtn').on('tap', function() {
    Ext.Msg.confirm("Confirmation", "Are you sure you want
    to delete this picture?", function(buttonId, value,
    opt) {
        if (buttonId == 'yes') {
            self.deletePicture(marker);
            popup.hide();
        }
    });
});

Ext.getCmp('closePhotoBtn').on('tap', function() {
    popup.hide();
});
});
}
```

We registered for event notifications using the addListener() event handler. This method takes an object, an event to listen for, and a function to call when the click event occurs. In the scope, this refers to the current marker. We will use it to access the picture property stored in the marker object. Inside the event handler, we instantiated the Travelly.view.Picture view, added it to the view port, and showed it on-screen. Similarly, we destroyed the popup on the hide event. Then, we assigned store in picture file URI to the image component and set the picture caption as text in the title bar. That's it. The following screenshot captures this discussion aptly:

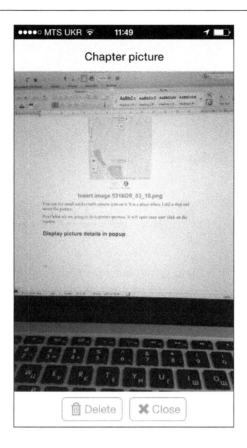

Now, we only need to click on the **Close** and **Delete** buttons.

When a user clicks on the closePhotoBtn button, we simply hide the popup.

When a user clicks on the deletePhotoBtn button, we display a delete confirmation message. If the user approves the deletion, we call the self.deletePicture function and pass the selected marker. We define this function in the Places controller:

```
deletePicture: function(marker) {
    var fileURI = marker.picture.get('url');
    fileURI = '/Travelly/' + fileURI.substring(fileURI.
lastIndexOf('/')+1);
    window.requestFileSystem(LocalFileSystem.PERSISTENT, 0,
    function(fileSys) {
```

```
        fileSys.root.getFile(fileURI, {create: false, exclusive:
        false}, function(fileEntry) {
            fileEntry.remove(onSuccess, onError);
        }, onError);
    },
    onError);

    var onSuccess = function(entry) {
        console.log("Removal succeeded");
    }

    var onError = function(error) {
        console.log("Error removing file: " + error.code);
    }

    var pictureStore = Ext.getStore('Pictures');
    pictureStore.remove(marker.picture);

    marker.setMap(null);
}
```

Here are the steps we use:

1. Remove the picture from persistent file storage with `fileEntry.remove`.
2. Remove the picture's metadata from the database with `pictureStore. remove(marker.picture)`.
3. Remove the marker from the map by calling the `setMap()` method and passing `null` as the argument.

Summary

In this chapter, you learned how to work with Cordova plugins. We used camera, geolocation, and filesystem plugins and several new Sencha Touch views and controllers to implement the simple "Travelly" application. Cordova played an interesting and important part in this small project. It handled the communication between the device and JavaScript part of the application well.

The next chapter is dedicated to integration with custom services. You will learn how to use Node.js to build RESTful services and how to communicate with it from our "Travelly" application.

4
Integrating the Travelly Application with Custom Service

In the previous chapter, you saw how to build a Travelly PhoneGap application using Sencha Touch and several PhoneGap plugins. You learned about the most popular architectural concept and checked out the available Sencha Touch components. This chapter is about building a REST API with Node.js and integrating it with our Travelly PhoneGap mobile application. We will implement routing mechanisms, request processing, and proper response sending.

In this chapter, we will cover the following topics:

- Basics of Node.js and its Express framework
- Writing a REST API with Express
- Developing service authentication and handling it on the application side
- Implementing MongoDB integration
- Coding the mobile application side with Sencha Touch and PhoneGap
- Displaying application's data on the website page

Discovering the REST API

Very often, we need to get data from a publicly available Internet service, send it over the Internet, and delete or modify it. We can do it all with REST APIs. We can have different URIs for different business models. Let's imagine that we need to make our products available in an Internet shop. To retrieve information about a specific product, we will perform a GET request to the /product/1 address, where 1 is the ID of the product. We can delete a product by making the DELETE request to the same resource. Similarly, we can use the PUT request to update the product and the POST request to create the product. The GET, POST, PUT, and DELETE methods are very often called verbs. In brief, the REST API consists of:

- HTTP methods:
  - **GET**: This is mainly used to retrieve data. It sends parameters in headers
  - **PUT**: This is used mainly to update a remote resource
  - **DELETE**: This is used to delete a resource
  - **PATCH**: This is used to update partial resources
  - **POST**: This submits data to be processed
  - Other request types

- **Base URI**: `http://mytravelly.com`
- **URL path**: `/pictures/12`
- **Media type**: HTML, JSON, XML, Atom and, so on.

According to Wikipedia, an **Application Programming Interface (API)** specifies how some software components should interact with each other. The API is usually the part of our program that is visible to the outside world. In this chapter, we will build one. It's an API of our pictures library that we get with our mobile device. The resources are the pictures, and they will be accessed through a REST API.

Now, let's choose the technologies for the implementation of our service.

Exploring technologies to build a REST API

The shortest way for us, as web developers, to build a REST API is to use a language we already know, JavaScript. Based on the selection, it is not hard to guess that we need to use Node.js and MongoDB. We already installed Node.js in *Chapter 1*, *Installing and Configuring PhoneGap*. Let's now take a closer look at what it is and how we can use it for our needs.

Understanding Node.js

The main advantages of Node.js are as follows:

- It does everything asynchronously, because it uses an event-driven approach.
- It uses the very fast JavaScript V8 engine.
- It shows very good performance, especially on scalable architectures.
- Same JavaScript code can be used on both client and server sides. For example, we can use the same JSON structures and the same validation code on clients and servers.

If you write a lot of code, sooner or later, you start realizing that your logic should be split into different modules. In most languages, this is done via classes, packages, or an other language-specific syntax. However, in JavaScript, we don't have classes natively. Everything is an object, and in practice objects inherit other objects. There are several ways to achieve object-oriented programming within JavaScript. You may use prototype inheritance, object literals, or play with function calls. Thankfully, Node.js has a standardized way of defining modules. This is approached by implementing CommonJS, which is a project specifying an ecosystem for JavaScript outside the browser.

In the chapter, we will install and use several Node.js modules to better understand how they work.

Introducing MongoDB

To build a REST API, we need a database that will store picture metadata. MongoDB is a NoSQL type of database. According to the Wikipedia entry on NoSQL databases:

> NoSQL databases store data in different format from relational databases (documents, key-value pairs etc.)

In other words, it's simpler than a SQL database and, very often, it stores information with the key-value type. Usually, such solutions are used when handling and storing large amounts of data. It is also a very popular approach when developing real-time applications.

MongoDB uses **JavaScript Object Notation (JSON)** to store data, and Node.js easily works with JSON. So, MongoDB and Node.js are often used together. JSON is very popular as a data transfer format for Web APIs.

Installing MongoDB with Homebrew

In *Chapter 1*, *Installing and Configuring PhoneGap*, we already installed Homebrew. So, to install the latest MongoDB version, we simply need to execute the following command:

```
$ brew install mongodb
```

We can check whether MongoDB is installed properly by running the mongod server or checking the MongoDB shell version command:

```
$ mongo --version
MongoDB shell version: 2.6.5
```

Now, we are ready to implement some code to work with MongoDB.

> To be able to connect to MongoDB from our REST API application, we should have the MongoDB daemon running. We will start MongoDB simply by running the mongod command in the terminal.

Developing a REST API

Let's plan our first REST API. With Express, it is very easy and the development itself does not really differ from website development.

Here is a summary of what we want to implement as a REST API:

Resource (URI)	POST (create)	GET (read)	PUT (update)	DELETE (destroy)
/pictures	Create new picture	List pictures	N/A (update all)	N/A (destroy all)
/pictures/1	Error	Show picture ID 1	Update picture ID 1	Destroy picture ID 1

- The format is JSON
- Bulk updates and bulk destroys are not safe, so we will not implement those
- POST, GET, PUT, DELETE equals to CREATE, READ, UPDATE, DELETE, with the abbreviation CRUD

Using Express

Building a REST API is pretty popular in the Node.js community. There are a lot of different ways to handle such a task. There are even ready-to-use modules such as rest.js or restify. However, we will use another framework called Express.

 Express is a web application framework for Node.js. With Express, we can build websites as well as web APIs.

Express is suitable for simple and complex applications because of its architecture. You can use some of the popular middleware, or you can add a lot of features and still keep things modular.

In general, everything in Node.js is doing two things: running a server that listens on a specific port and processing incoming requests. You may think of Express as a wrapper of these two functionalities. Here is an example from the official documentation on Node.js:

```
var http = require('http');
http.createServer(function (req, res) {
    res.writeHead(200, {'Content-Type': 'text/plain'});
    res.end('Hello World\n');
}).listen(1337, '127.0.0.1');
console.log('Server running at http://127.0.0.1:1337/');
```

Where:

- `req` is an instance of `http.IncomingMessage`, and we can get different request parameters
- `res` is an instance of `http.ServerResponse`, and we can manipulate the response in different ways

As you can see, we are using the `http` native module and running the application on port `1337`. There is also a request handler function that simply sends the `Hello world` string to the browser. On Express, it will look like this:

```
var express = require('express');
var app = express();
app.get("/", function(req, res, next) {
    res.send("Hello world");
}).listen(1337);
console.log('Server running at http://127.0.0.1:1337/');
```

We don't need to specify the response headers or add a new line at the end of the string. That's because the framework does it for us. Along with that, we have a bunch of middleware that will help us to process the requests easily. With Express, we have a lot of tools that do the boring stuff, so we can focus on the application's logic and content. That's what Express is built for, saving time for the developer by providing ready-to-use functionality.

We selected Express exactly because of its ability to create both web pages and REST APIs.

First of all, let's install Express:

```
$ npm install -g express
```

Also, we need to install express-generator:

```
$ npm install -g express-generator
```

The `express-generator` command has a set of command-line tools to generate an Express application.

We want to install an Express npm module globally. So, we will add the `-g` flag:

- If we want some tool available only from a project's folder, we would run the npm command without the `-g` flag. This means the module will be installed locally.

- If we install a tool with the `-g` flag, it would become available from different places because its binaries are added in the PATH environment variable.

We can check the version of the installed Express application to validate that we have successfully installed it:

```
$ express --version
4.2.0
```

You can see that, in my case, I got Express version `4.2.0` installed. Now, we can move forward and create our first Express application.

Generating an Express application

There are two possible ways to create an Express application: manually or with the Express application generator. The application generator tool quickly creates an application skeleton. Let's take the short path and use the generator. First of all, let's install it:

```
$ npm install express-generator -g
```

Let's go into the desired folder and run the following command:

```
$ express -e travelly-svc
```

This command consists of the following elements:

- `-e` add EJS engine support

 EJS is an open source JavaScript template library. EJS combines data and a template to produce HTML.

- `travelly-svc` is the application's name

This command creates an app named `travelly-svc` in the current working directory.

After that, install dependencies:

```
$ cd travelly-svc
$ npm install
```

> This command installs a package, and any packages that it depends on.

The `travelly-svc` folder contains a program described by a `package.json` file with the following content:

```
{
    "name": "travelly-svc",
    "version": "0.0.1",
    "private": true,
    "scripts": {
        "start": "node ./bin/www"
    },
    "dependencies": {
        "express": "~4.2.0",
        "static-favicon": "~1.0.0",
        "morgan": "~1.0.0",
        "cookie-parser": "~1.0.1",
        "body-parser": "~1.0.0",
        "debug": "~0.7.4",
        "ejs": "~0.8.5"
    }
}
```

The `npm install` command installs all modules in the `node_modules` folder. In our case, the Express generator has several dependencies: Express framework, the `static-favicon` middleware to serve favicons from the `/favicon.ico` route, the `morgan` logger middleware, `cookie-parser` to parse cookie headers into `req.cookies`, `body-parser` to parse the body into `req.body`, the `debug` tool, and the `ejs` template engine.

At the end of the `npm install` command execution, you should see something like this:

```
debug@0.7.4 node_modules/debug

static-favicon@1.0.2 node_modules/static-favicon

ejs@0.8.8 node_modules/ejs

morgan@1.0.1 node_modules/morgan
└── bytes@0.3.0

cookie-parser@1.0.1 node_modules/cookie-parser
├── cookie-signature@1.0.3
└── cookie@0.1.0

body-parser@1.0.2 node_modules/body-parser
├── qs@0.6.6
├── raw-body@1.1.7 (string_decoder@0.10.31, bytes@1.0.0)
└── type-is@1.1.0 (mime@1.2.11)

express@4.2.0 node_modules/express
├── parseurl@1.0.1
├── utils-merge@1.0.0
├── merge-descriptors@0.0.2
├── cookie@0.1.2
├── escape-html@1.0.1
├── debug@0.8.1
├── cookie-signature@1.0.3
├── fresh@0.2.2
├── qs@0.6.6
├── range-parser@1.0.0
├── methods@1.0.0
├── buffer-crc32@0.2.1
├── serve-static@1.1.0
```

```
├── path-to-regexp@0.1.2
├── type-is@1.1.0 (mime@1.2.11)
├── send@0.3.0 (debug@0.8.0, mime@1.2.11)
└── accepts@1.0.1 (negotiator@0.4.9, mime@1.2.11)
```

Now, we are ready to use Express. If you type `require('express')` in the Node.js console, it will start looking for that library inside the local `node_modules` directory. If you miss the running of `npm install`, you will probably get `Error: Cannot find module 'express'`.

To run the application, we can execute following command:

```
$ DEBUG=travelly-svc ./bin/www
```

Now, in the browser, we can open `http://localhost:3000/` to access our Express application.

The generated `app` directory structure looks like this:

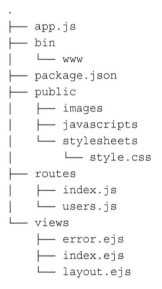

```
.
├── app.js
├── bin
│   └── www
├── package.json
├── public
│   ├── images
│   ├── javascripts
│   └── stylesheets
│       └── style.css
├── routes
│   ├── index.js
│   └── users.js
└── views
    ├── error.ejs
    ├── index.ejs
    └── layout.ejs
```

We can use a different folder structure but, for the time being, I would like to follow the proposed default structure to avoid complexity in the beginning. A few components of the preceding directory are explained here:

- `bin/www`: This is the entry point of the application. It creates an HTTP server listening on the 3000 port. This file requires `app.js`.

- `app.js`: This is the application's main file. It contains all the main configurations and requirements.

- `package.json`: This contains the program description with its dependencies.
- `public`: This is publicly accessible folder. It contains JavaScript, styles, and pictures.
- `routes`: Here is our main business logic of the application.

 If you are familiar with the model-view-controller pattern, these are the controllers of our application.

- `views`: Here is all the UI (`views`). In our case, we are using the EJS template engine.

Exploring the basic Express application

We generated a basic Express application, but we do not yet know how it works. Let's look at the generated application in detail.

The `app.js` file starts with initializing the module dependencies:

```
var express = require('express');
var path = require('path');
var favicon = require('static-favicon');
var logger = require('morgan');
var cookieParser = require('cookie-parser');
var bodyParser = require('body-parser');
```

This is exactly what is defined as a dependency in `package.json`. We already explained what each module means, so let's move forward with the following script:

```
var routes = require('./routes/index');
var users = require('./routes/users');
```

These are the controllers for our application. We will add controllers here when we need them.

Just after that, an `app` variable is created. It represents the Express library:

```
app = express();
```

We use that variable to configure our application. The script continues with setting some key-value pairs:

```
app.set('views', path.join(__dirname, 'views'));
app.set('view engine', 'ejs');
```

We configure the path to our views files and instruct Express to use EJS as a template engine. The next few lines of code add middleware to the framework:

```
app.use(favicon());
app.use(logger('dev'));
app.use(bodyParser({limit: '50mb'}));
app.use(bodyParser.json());
app.use(bodyParser.urlencoded({ extended: true }));
app.use(cookieParser());
app.use(express.static(path.join(__dirname, 'public')));
```

The first middleware is for feeding favicon to the client. The second one is responsible for the output in our console. If you remove it, you would not get information about the incoming requests to your server. Here is a simple output produced by the logger:

```
GET / 304 5ms
GET /stylesheets/style.css 304 1ms
```

json and urlencode are related to the data sent along with the request. We need them because they convert the information in an easy-to-use format. There is also middleware for the cookies. They populate the request object so that we have access to the needed data. At the end, we defined our static resources that should be delivered by the server:

 Middlewares is a stack of functions that runs on each request to the server, one after another.

```
app.use('/', routes);
app.use('/users', users);
```

These preceding lines are held responsible by the URL for controller mapping. There are two routes initially. These routes are just an example of how we could eventually write our controllers.

These are just few lines, but in them we've configured the whole application. We may remove or replace some of the modules, and the others will continue to work. The script goes on to enable error handling:

```
if (app.get('env') === 'development') {
    app.use(function(err, req, res, next) {
        res.status(err.status || 500);
        res.render('error', {
```

```
                message: err.message,
                error: err
            });
        });
    }
```

As you can see, this is done only in the development mode. It's a good practice to have a development server where you can test the application one more time before going into a production state.

Now, if we look at the `bin/www` file, we will find code that it creates a HTTP server and listens on `3000` port:

```
app.set('port', process.env.PORT || 3000);
var server = app.listen(app.get('port'), function() {
  debug('Express server listening on port ' + server.address().port);
});
```

Handling URLs with routes

The input of our application is the routes. The user visits our page on a specific URL, and we have to map this URL to a specific logic. In the context of Express, this is easy to do. Let's create a new route in the `routes/pictures.js` file with the following content:

```
var express = require('express');
var router = express.Router();
```

A `router` object is an isolated instance of middleware and routes.

To handle a specific URL, we can add the following lines to our route file:

```
router.get('/', function(req, res, next) {
  console.log('Retrieve and display pictures...');
});
```

Now, we need to include this route in `app.js`:

```
var pictures = require('./routes/pictures');
app.use('/pictures', pictures);
```

That is it! From now on, our application will handle any request that ends in `/ pictures`.

If you open your browser and type `http://localhost:3000/pictures` in the terminal, you will see **Retrieve and display pictures...**.

We even have control on the HTTP's method, that is, we are able to catch POST, PUT, or DELETE requests. This is very handy if you want to keep the address path, but apply a different logic. Here is an example:

```
router.post('/', function(req, res, next) {
    // ...
});
router.get('/:id', function(req, res, next) {
    // ...
});
router.put('/:id', function(req, res, next) {
    // ...
});
router.delete('/:id', function(req, res, next) {
    // ...
});
```

The path is still the same, /pictures, but if you make a POST request to that URL, the application will try to create a new picture. Otherwise, if the method is GET, it will return a list of all the pictures. There is also the app.all method that you can use to handle all the method types at once.

Returning a response

After our server accepts a request and does some stuff, it should send a response to the client. This could be HTML, JSON, XML, binary data, or something else. By default, each middleware in Express accepts request and response. The response object has a few methods that take care of the answer of the request object. In practice, it is not simple. For example, every response should have a proper content type or content length. Express simplifies the setting of these headers and makes our life easier. We will use the .send, .json, and .end methods, shown as follows:

- res.send sends the response where the body parameter can be a buffer object, a String, an object, or an Array. Here is an example:

  ```
  res.send('<p>some html</p>');
  ```

- res.json([body]) sends a JSON response. Here is an example:

  ```
  res.json({ user: 'Andrew' });
  ```

- res.end([data] [, encoding]) is used to quickly end the response without any data. Here is an example:

  ```
  res.end();
  ```

When we pass a string, the framework sets the `Content-Type` header to `text/html`. If we develop an API, the response status code is going to be important for us. With Express, we are able to set it like this:

```
res.status(404).end();
```

While we build websites or applications with a user interface, we normally need to serve HTML. It's a good practice to use a template engine. We save everything into external files, and the engine reads the markup from there. It probably populates it with some data and, at the end, provides a ready-to-show template. You may be surprised but, in Express, the whole operation is summarized in only one method: `.render`. However, to work properly, we have to instruct the framework about the type of template engine to use. If you remember, we already talked about this at the beginning of this chapter:

```
app.set('views', path.join(__dirname, 'views'));
app.set('view engine', 'ejs');
```

The first line sets the path to the EJS files, and the second one defines the engine. If we open `views/index.ejs`, we might be able to see these lines in the `body` section:

```
<h1><%= title %></h1>
<p>Welcome to <%= title %></p>
```

Express provides a method for serving templates. It accepts the path to the template, data to be applied, and a callback. To render the preceding template, we should use:

```
res.render("index", {title: "Travelly page title"});
```

The generated HTML looks like this:

```
<h1>Travelly page title </h1><p>Welcome to Travelly page title</p>
```

Connecting Express and MongoDB

As we discussed earlier, we need to store pictures' metadata somewhere for later access. Hopefully, you have already installed MongoDB, and we can start it by typing `mongod` in terminal.

To access MongoDB from the Node.js application, let's install the `mongoose` module:

```
$ npm install mongoose --save
```

 The `--save` option helps us save the package inside the dependencies section of `package.json`.

With mongoose, we can define business data models, configure validation, and so on.

Now, we will include mongoose in app.js:

```
var mongoose = require('mongoose');
mongoose.connect('mongodb://localhost/travelly', function(err) {
    if(err) {
        console.log('connection error', err);
    } else {
        console.log('connection successful');
    }
});
```

Now, when we run DEBUG=travelly-svc ./bin/www, we will notice the message, connection successful. We have a pending connection to the travelly database running on localhost.

Creating a picture model

We have installed all the required modules for a basic API. We can now start implementing it.

Let's create the models directory with Picture.js in it:

```
var mongoose = require('mongoose');
var PictureSchema = new mongoose.Schema({
    url: String,
    title: String,
    lon: String,
    lat: String,
    fileName: String,
    created_at: { type: Date, default: Date.now }
});
module.exports = mongoose.model('Picture', PictureSchema);
```

As you can see, we have defined the `String` and `Date` types for the fields in `Schema`. With mongoose, we can use the following types:

- Boolean
- Number
- String
- Date
- Array
- ObjectId
- Mixed
- Buffer

`mongoose.model` is compiling our schema into a model.

 Models are fancy constructors compiled from our schema definitions. Instances of these models represent documents that can be saved and retrieved from our database. All document creation and retrieval from the database is handled by these models.

Now, let's come back to our `route/pictures.js` file and modify it a little. At the top of the file, add the following lines:

```
var mongoose = require('mongoose');
var Picture = require('../models/Picture.js');
```

Here, we include `mongoose` and our created model. Now we are going to modify the existing code to retrieve pictures from the database:

```
router.get('/', function(req, res) {
    Picture.find(function(err, pictures) {
        if (err) return next(err);
        res.json(pictures);
    });
});
```

Here, `Picture.find` executes MongoDB's find function to retrieve all picture documents from the database. If we get an error, we continue with the error. However, if we get success, we can send the response to the client in the JSON format with all the pictures found.

Right now, we do not have any picture in the database, but we can verify that it is working. You can open `http://localhost:3000/pictures` in the browser, but it is limited. For test purposes, I would recommend you to use the Postman Chrome plugin (`https://chrome.google.com/webstore/detail/postman-rest-client/fdmmgilgnpjigdojojpjoooidkmcomcm?hl=en`). It allows us to use all the HTTP verbs easily and check `x-www-form-urlencoded` to add parameters. For example, here is the screen of the GET request to `http://localhost:3000/pictures`:

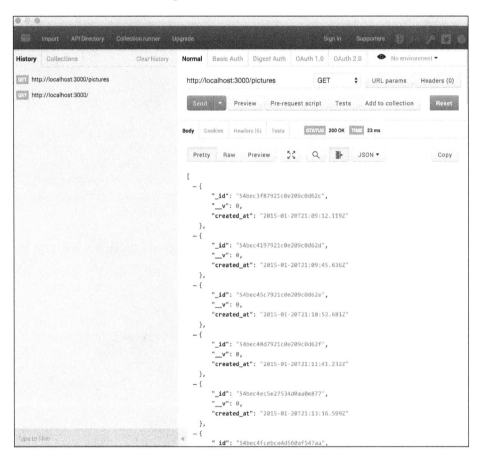

Similarly, we can implement rest of the CRUD operations.

Creating a new picture record

The addition of a new picture into the database should happen via the POST request. Here is the route that will handle this task:

```
router.post('/', function(req, res, next) {
  var picture = req.body;
    Picture.create(picture, function(err, post) {
        if (err) return next(err);
        res.json(post);
    });
});
```

Now, this POST request handler will take the request data and create a new picture record. We pass JSON data to the `/pictures` URL.

Editing a picture record

To update the picture model, we need to implement the PUT request handler. It should handle the picture ID as well:

```
router.put('/:id', function(req, res, next) {
    Picture.findByIdAndUpdate(req.params.id, req.body,
    function(err, post) {
        if (err) return next(err);
        res.json(post);
    });
});
```

Test in Postman using the `_id` parameter from which you created elements; for example, `http://localhost:3000/pictures/54bec3f87921c0e209c0d62c`.

Deleting a record

Deleting records is really similar to editing them. Again, we need a dynamic route:

```
router.delete('/:id', function(req, res, next) {
    Picture.findByIdAndRemove(req.params.id, req.body,
    function(err, post) {
        if (err) return next(err);
        res.json(post);
    });
});
```

As you can see, if we get an error, we pass it to the client as well.

Implementing service authentication

Now, when we have some basic functionality implemented, we would like to add some security to our REST API. The main difference between a web application and web API lies in session and cookie usage. In the API, we should not use them at all, but we have to implement some security. We can implement security with the **JSON Web Tokens (JWTs)** mechanism.

 You can learn more about JWT from its specification `http://tools.ietf.org/html/draft-ietf-oauth-json-web-token-19`.

With this approach, we should not send the username and password with each request. The approach provides a mechanism to create a secure token and pass it with every request. A token is different from a password, and it is much longer. Usually, it contains several information groups encoded with a hashing algorithm.

A JWT consists of three parts, as follows:

- The token and hashing algorithm:

```
{
  "typ" : "JWT",
  "alg" : "HS256"
}
// after encoding: eyJ0eXAiOiJKV1QiLCJhbGciOiJIUzI1NiJ9
```

- JWT Claims Set , which has `issuer`, which specifies the person making the request, and `expires`, which limits the lifetime of the token:

```
{
  "iss": "admin",
  "exp": 1300819560
}
// after encoding:  eyJpc3MiOiI1NGM1NWFhYzE0YjU4MTdlNGYwZWQzND
QiLCJleHAiOjE0MjM2OTkxOTAwMDF9
```

- A signature generated based on the header and the body:

```
tguxtj0UqtQYvFmT572Sv1GRemSG8qZ6pPT_0fgRlv0
```

The resulting complete JWT looks like this:

```
eyJ0eXAiOiJKV1QiLCJhbGciOiJIUzI1NiJ9.eyJpc3MiOiI1NGM1NWFhYzE0Yj
U4MTdlNGYwZWQzNDQiLCJleHAiOjE0MjM2OTkxOTAwMDF9.tguxtj0UqtQYvFmT5
72Sv1GRemSG8qZ6pPT_0fgRlv0
```

So, you do not need to implement all this from scratch. There is a ready-to-use Node.
js `jwt-simple` module for it.

Let's install it as we already did for the `mongoose` module earlier:

$ npm install jwt-simple --save

In addition, we will need a few other modules:

- `bcrypt` is a library to help us hash passwords
- `moment` is a JavaScript date library to parse, validate, manipulate, and format dates

Now, let's install them:

$ npm install bcrypt --save

$ npm install moment --save

In `app.js`, add a secret for encoding and decoding JWT tokens:

```
app.set('jwtTokenSecret', 'secret-value')
```

Another thing we need to do is enable the clients to exchange their username and password for a token. In this case, we will implement a login form, which will send a POST request to the authentication endpoint.

Implementing a login form

We want our login form to look nice. So, we will use Twitter Bootstrap (`http://getbootstrap.com/`) and its Flatly theme (`http://bootswatch.com/flatly/`). First of all, download the Bootstrap package and unpack it under `public/javascripts/lib/`. After that, simply download a CSS file from `http://bootswatch.com/flatly/` and replace the one in Bootstrap.

Now, we need to create a proper route and view for the login form. Let's create a view under `views/login.ejs` with Bootstrap included in the `head` section and the following content in the `body` section:

```
<div class="panel-body">
  <%if (message) { %>
  <div class="alert alert-danger" role="alert"><%= message
  %></div>
  <% } %>
  <form id="loginForm" action="/login" method="post">
    <fieldset>
      <div class="form-group">
        <input class="form-control" placeholder="Username"
        name="username" type="text">
      </div>
      <div class="form-group">
        <input class="form-control" placeholder="Password"
        name="password" type="password" value="">
      </div>
      <input class="btn btn-lg btn-success btn-block"
      type="submit" value="Login">
    </fieldset>
  </form>
</div>
```

Here, the tag with the `class` alert is a placeholder to display error messages about problems with authentication. You can see that we described here a form with the POST action to the `/login` route. The form contains two fields: `username` and `password`. These fields' data will be sent when the form is submitted.

Let's create route under `routes/auth.js` with the following content:

```
var express = require('express');
var router = express.Router();
router.get('/', function(req, res) {
  res.render('login', { title: 'Login', message: null });
});
```

Do not forget to include this route in `app.js`:

```
var auth = require('./routes/auth');
app.use('/login', auth);
```

Now, if you open `http://localhost:3000/login` in the browser, you will see something like this:

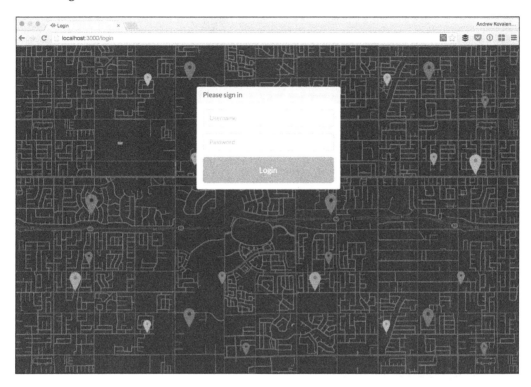

Handling the authentication endpoint request

To be able to provide authentication, we need to store users and their passwords somewhere. We already successfully used MongoDB and Mongoose. Let's create a user schema and model. Under `models/User.js`, add the following lines of code:

```
var mongoose = require('mongoose');
var UserSchema = mongoose.Schema({
    username: { type: String, required: true, index: { unique:
    true } },
    password: { type: String, required: true }
});
module.exports = mongoose.model('User', UserSchema);
```

Here, we created `UserSchema` with two fields: `username` and `password`. These are the same fields as on the login form.

With Mongoose, we can define static methods, which are very handy:

```
UserSchema.statics.findByUsername = function (username, cb) {
    this.findOne({ username: username }, cb);
}
```

We will use this function in the authentication route. This method will search for the user document on the database based on its username.

Now, we will implement a password-encoding process with the bcrypt library:

```
var bcrypt = require('bcrypt');
var SALT_WORK_FACTOR = 10;
UserSchema.pre('save', function(next) {
    var user = this;
    if (!user.isModified('password')) return next();
    bcrypt.genSalt(SALT_WORK_FACTOR, function(err, salt) {
        if (err) return next(err);
        bcrypt.hash(user.password, salt, function(err, hash) {
            if (err) return next(err);
            user.password = hash;
            next();
        });
    });
});
```

Here, we used the pre middleware to intercept the User object, generate salt and hash, and continue the save process.

To compare the password in the database and the password sent from the login form, we need another method:

```
UserSchema.methods.comparePassword = function(candidatePassword, cb) {
    bcrypt.compare(candidatePassword, this.password, function(err,
    isMatch) {
        if (err) return cb(err);
        cb(null, isMatch);
    });
};
```

We defined the custom document instance method in the preceding code.

Now, let's add a handler to the authentication POST request to `routes/auth.js`:

```
router.post('/', function(req, res) {
    if (req.body.username && req.body.password) {
        UserModel.findOne({
            username: req.body.username
        }, function(err, user) {
            if (err || !user) {
            // authentication failed
            }
            user.comparePassword(req.body.password, function(err,
            isMatch) {
                if (err) {
                    // authentication failed
                }
                if (isMatch) {
                    // authentication success
                } else {
                    // authentication failed
                }
            });
        });
    } else {
        // authentication failed
    }
});
```

Here, before returning a successful response with the token, our code does the following validations:

- Check whether the username and password are provided
- Find the appropriate user by username to verify that it exists
- Check whether the password is correct

If the authentication fails, we could simply display the same login form with the following error message:

```
res.render('login', { title: 'Login', message: 'Authentication failed.
Login or password is incorrect.' });
```

With a successful response, we can send the JWT token in the URL:

```
var expires = moment().add('days', 7).valueOf()
var token = jwt.encode({
        iss: user.id,
        exp: expires
    },
    app.get('jwtTokenSecret')
);
res.redirect('/?token=' + token);
```

There are two parameters in the `jwt.encode()` function, as follows:

- The body of the token (we used the `Moment.js` library to set expiration to 7 days from now)
- The secret string we defined in `app.js`

The `res.redirect()` method is used to return the token to the client.

Verifying authentication

Now, we have to verify that we get the authenticated request. We will add a method that performs the following steps:

1. Check the token
2. Decode the token and verify it
3. If the token is valid, get the user from the MongoDB and attach it to the request

We will create a basic structure of the preceding steps in `lib/jwtauth.js`:

```
var url = require('url')
var UserModel = require('../models/User.js')
var jwt = require('jwt-simple');

module.exports = function(req, res, next){
  // code goes here
}
```

You can see that here, we required an additional `url` module.

This module has utilities for URL resolution and parsing, intended to ensure feature parity with the Node.js core `url` module.

With the help of this module, we will parse the token from the URL:

```
var parsed_url = url.parse(req.url, true);
var token = (req.body && req.body.access_token) || parsed_url.query.
access_token || req.headers["x-access-token"];
```

You can see that we are checking `access_token` in different places: URL, body, and header. So, this means that the client can send it in different ways.

Next, we need to decode JWT:

```
if (token) {
  try {
    var decoded = jwt.decode(token, app.get('jwtTokenSecret'))
      // work with decoded token here
  } catch (err) {
    return next()
  }
} else {
  next()
}
```

If the decode process fails, it will throw an exception. If there is no token, we could simply continue processing the request. If the token exists and is decoded, we should check whether it has expired:

```
if (decoded.exp <= Date.now()) {
  res.end('Access token has expired', 400)
}
```

If the code is valid, we can get the user now and attach it to the request:

```
UserModel.findOne({ '_id': decoded.iss }, function(err, user){
  if (!err) {
    req.user = user
    return next()
  }
})
```

Now, we need to attach the middleware to a route. Let's add the following lines of code to the `routes/pictures.js` route:

```
var jwtauth = require('../lib/jwtauth');
router.get('/', jwtauth, function(req, res) {
  // code goes here
});
```

We can add the `jwtauth` middleware to any route.

Our middleware now examines requests looking for a valid token and, if one exists, attaches a user object to the request. There is something interesting about the routing in Express. You can pass not only one, but many handlers. This means that you could create a chain of functions corresponding to one URL. So, let's add another simple middleware to check the `req.user` object:

```
var requireAuth = function(req, res, next) {
    if (!req.user) {
        res.status(401).end('Not authorized')
    } else {
        next()
    }
}
```

Here, we send the `Not authorized` response if there is no user object passed from the `jwtauth` middleware. The last thing we need to do about authentication is add this new middleware to the route right after `jwtauth`:

```
router.get('/', jwtauth, requireAuth, function(req, res) {
    // code goes here...
});
```

That is it! We successfully implemented authentication on the REST API side. Now, we can create a dummy user in the database. We can do this with the help of a script we place in the `seed.js` file:

```
var mongoose = require('mongoose');
var UserModel = require('./models/User')
mongoose.connect('mongodb://localhost/travelly');
var db = mongoose.connection;
db.on('error', console.error.bind(console, 'Failed to connect to
database!'))
db.once('open', function callback () {
  var user = new UserModel()
  user.username = 'admin'
  user.password = 'password'
  user.save(function(err){
    if (err) {
      console.log('Could not save user.')
    } else {
      console.log('Database seeded')
    }
    process.exit()
  })
});
```

In the script, we connect to the database and user with the username as `admin` and password as `password`. We simply need to execute this script to add the user to the database:

```
$ node seed.js
```

Implementing authentication on the application side

Now, let's go back to our Cordova/Sencha Touch application. We need to be able to authenticate in the mobile application, receive the token, save it in the application, and use it for all future requests. We already have the login form implemented on the service side. We can utilize it in our mobile application as well.

We can easily handle this task with the `InAppBrowser` Cordova plugin. This plugin provides a web browser view that displays when calling `window.open()`. Let's install the plugin:

```
$ cordova plugin add org.apache.cordova.inappbrowser
```

Now, let's define the UI from where we can invoke `InAppBrowser`. We can do this in `app/view/Main.js`. We will create the additional `Settings` tab panel:

```
{
    title: 'Settings',
    iconCls: 'settings',
    items: [
        {
            docked: 'top',
            xtype: 'titlebar',
            title: 'Settings'
        },
        {
            xtype: 'button',
            id: 'loginBtn',
            text: 'Login',
            iconCls: 'user',
            margin: 5
        },
        {
            xtype: 'button',
            id: 'logoutBtn',
            text: 'Logout',
            iconCls: 'user',
```

```
              margin: 5,
              hidden: true
          }
      ]
  }
```

This panel contains the login and logout buttons. By default, the logout button is hidden. We will click on the login button when we want to open InAppBrowser with our login form. Let's handle this in the app/controller/Main.js controller:

```
doLogin: function() {
    var self = this;
    var svcUrl = Travelly.app.getGlobal('svcUrl');
    var authWindow = window.open(svcUrl + '/login', '_blank',
    'location=yes');
}
```

Here, you can see that we try to open the svcUrl + '/login' URL. We retrieve svcUrl from the application's global variables list, which we define in app.js:

```
globals: {
    svcUrl: 'http://192.168.1.227:3000'
},
getGlobal: function(key)
{
    if (this.globals[key] !== undefined) {
        return this.globals[key];
    }
    return null;
}
```

The getGlobal is just a helper function to retrieve a property value by its name. You can see this in svcUrl; I have entered the local IP address of the computer my service is running on. Usually, my computer and mobile device are in the same local network, and I can easily test interaction between my application on the device and service. In the real world, I would need to replace svcUrl with a hosted online address, for example, http://mytravelly.com/api.

Now, when I run the latest version of the application on my device, go to the
Settings tab, and click on the **Login** button, I should see something like this:

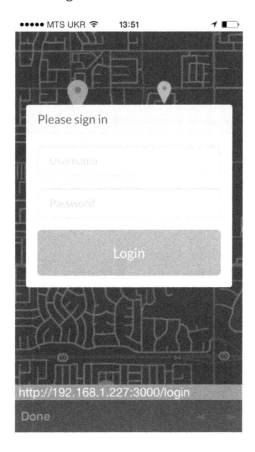

We can enter the username and password of the dummy user and hit **Login**.
If the authentication is successful, we will be redirected to the same page with
the JWT token passed—for example, `http://192.168.1.227:3000/login/`
`?token=eyJ0eXAiOiJKV1QiLCJhbGciOiJIUzI1NiJ9.eyJpc3MiOiI1NGM1NWFh`
`YzE0YjU4MTdlNGYwZWQzNDQiLCJleHAiOjE0MjM2OTkxOTAwMDF9.tguxtj0UqtQY`
`vFmT572Sv1GRemSG8qZ6pPT_0fgRlv0`.

We can handle it in our application on the same `doLogin` method:

```
doLogin: function() {
    // ...
    authWindow.addEventListener('loadstart', function(e) {
        var url = e.url;
        var token = self.getParam(url, 'token');
        if (token) {
            // save token code goes here
        }
    });
}
```

Here we are listening for the `loadstart` event. We parse the URL and look for the token parameter. If it is present, we need to store it in the application. Let's create the Sencha Touch store where we can save the token. We will implement it as key-value storage under `app/store/Settings.js`:

```
Ext.define('Travelly.store.Settings', {
    extend: 'Ext.data.Store',
    requires: ['Travelly.model.Settings'],
    config: {
        storeId: 'Settings',
        model: 'Travelly.model.Settings',
        autoLoad: true,
        autoSync: true
    }
});
```

The attached module looks like this:

```
Ext.define('Travelly.model.Settings', {
    extend: 'Ext.data.Model',
    config: {
        fields: [
            { name: 'id', type: 'int' },
            { name: 'name', type: 'string' },
            { name: 'value', type: 'string' }
        ]
    }
});
```

It has only three fields: id, name, and value. For easier access to the settings, let's add helper set and get methods to the store:

```
get: function(name)
{
    var settingsStore = this,
        setting = settingsStore.findRecord('name', name),
        value = null;
    if (setting) {
        value = setting.get('value');
    }
    return value;
},
set: function(name, value)
{
    var store = this,
        record = store.findRecord('name', name);
    if (! record) {
        record = Ext.create('Travelly.model.Settings');
        record.set('name', name);
        record.set('value', value);
        store.add(record);
    } else {
        record.set('value', value);
        record.save();
    }
}
```

In the get method, we find the setting by its name and return value. In the set method, we pass both name and value. If we find a setting with such a name, we update it; otherwise, we just create a new setting record. Again, for easier access to these functions, I will wrap them in the global scoped functions in app.js:

```
getSetting: function(name)
{
    return Ext.getStore('Settings').get(name);
},
setSetting: function(name, value)
{
    Ext.getStore('Settings').set(name, value);
}
```

It is pretty clear. Now, what we need is to come back to our authentication token handler and save our token as a setting:

```
Travelly.app.setSetting('token', token);
authWindow.close();
var loginBtn = self.getLoginBtn();
var logoutBtn = self.getLogoutBtn();
loginBtn.setHidden(true);
logoutBtn.setHidden(false);
```

Here, we saved the token, closed the authentication window, hid the login button, and displayed the logout button. Now, when we go to the **Setting** tab again, we will see the **Logout** button there. By clicking on this button, we simply clear the token value in settings, hide the logout button, and show the login button:

```
doLogout: function() {
    var self = this;
    Travelly.app.setSetting('token', '');
    var loginBtn = self.getLoginBtn();
    var logoutBtn = self.getLogoutBtn();
    loginBtn.setHidden(false);
    logoutBtn.setHidden(true);
}
```

Now, we have the token and can pass requests to our REST API. However, before doing it, let's go back to the service side and implement the file upload functionality. We need to store pictures along with their metadata.

Implementing file upload on the service side

There is a great library to handle file uploads for Node.js Express framework. It is a Multer module.

 Multer is a middleware that handles multipart/form-data.

Let's install it:

```
$npm install multer --save
```

The configuration of Multer is pretty straightforward. We will use it as a middleware after our authentication to be sure that only the authenticated request uploads files to the service. We will use the following code for this purpose:

```
var multer = require('multer');
router.post('/', jwtauth, requireAuth, multer({
    dest: './public/uploads/',
    rename: function(fieldname, filename) {
        return Date.now();
    }
}), function(req, res, next) {
  var picture = req.body;
  if (req.files && req.files && req.files.picture) {
      picture.fileName = req.files.picture.name;
  }
  // save to the DB code goes here
});
```

Where:

- `dest` is the destination directory for the uploaded files.
- `rename` is a customized function to assign new file names to avoid conflicts. In our case, we used timestamp.
- `req.files` is an object containing Multer file objects. In our cases we have only a single file object picture, where the name is the renamed file name.

We take the saved file name and save it to the database along with other metadata so that we can access the file by its name later.

Now, we are ready to receive files from the client.

Implementing file upload on the application side

To be able to upload files from our mobile Cordova application, we need to install an additional `FileTransfer` plugin:

```
$ cordova plugin add org.apache.cordova.file-transfer
```

 This plugin allows you to upload and download files. This plugin defines global `FileTransfer` and `FileUploadOptions` constructors.

We will now change the `savePhoto` method in the `Main.js` controller:

```
if (Travelly.app.getSetting('token')) {
    self.savePhotoToService(picture, function(isSuccess,
    errorMessage) {
        setTimeout(function(){
            popup.hide();
        },3000)
    });
} else {
    alert('Please login if you want to save picture online.');
}
```

Here, we checked whether we have the token stored in settings and called the `self.savePhotoToService` method if the token is present:

```
savePhotoToService: function(picture, callback) {
    var self = this,
        svcUrl = Travelly.app.getGlobal('svcUrl'),
        publishUrl = svcUrl + '/pictures?access_token=' +
        Travelly.app.getSetting('token'),
        fileTransfer = new FileTransfer(),
        fileOptions = new FileUploadOptions(),
        uploadParams = {
            title: picture.get('title'),
            lat: picture.get('lat'),
            lon: picture.get('lon')
        };
    // upload code goes here
}
```

Here, we will define the main variables needed for file uploading:

- `publishUrl` is a URL where we will upload our picture with metadata
- `fileTransfer` is an object that helps us upload files (a HTTP multipart POST request)
- `fileOptions` is an object with the following optional parameters:
  - `fileKey`: This is the name of the form element. It defaults to `file`.
  - `fileName`: This is the file name to use when saving the file on the server. It defaults to `image.jpg`.
  - `mimeType`: This is the mime type of the data to upload. It defaults to `image/jpeg`.
  - `params`: This is a set of optional key/value pairs to pass in the HTTP request.
- `uploadParams` is the metadata we will send with pictures

Now, we will assign the upload options:

```
var imageUrl = picture.get('url');
fileOptions.fileKey = 'picture';
fileOptions.fileName = imageUrl.substr(imageUrl.lastIndexOf('/') + 1);
fileOptions.mimeType = 'image/jpeg';
fileOptions.params = uploadParams;
```

Here, we already described `fileKey`, `fileName`, `mimeType`, and `params`.

As a filename, we take a substring from the full image path. We now have to invoke the upload method:

```
fileTransfer.upload(imageUrl, publishUrl, function(response) {
        if (response.responseCode === 200) {
            if (callback) callback(true);
        } else {
            if (callback) callback(false, 'Wrong status');
        }
    },
    function(error) {
        alert('Some error ocurred (' + error.body + '). Please try
        again later.');
    },
    fileOptions
);
```

The `fileTransfer.upload` method sends a file to the server with `fileOptions` attached to it. In the callback, we check the response code, and if it is successful, we would simply close the details window.

If we run our application, take a picture, and save it, the application will send it to the service. We can check pictures in the public/uploads folder on the service. In my case, the folder looks like this:

```
.
├── 1422364937548.jpg
└── 1422366001590.jpg
```

Now, we have data on the server, and we can work with it as we want. For example, we can build a map page with picture markers.

> You can check the `http://localhost:3000/map` page of the service in the files attached to the code of this chapter.

Summary

In this chapter, we built a REST API to store information about pictures. Node.js handles such tasks nicely, because it has native modules. That are easy to work with. We successfully covered GET, POST, PUT, and DELETE requests, creating an interface to manage pictures. In addition, we applied basic security practices to limit unauthorized access. You also learned how to work with the service from the client.

In the next chapter, we will build a HTML5 mobile game using HTML5 Canvas and its 2D context. You will learn how to build HTML5 animations, how to handle mobile gestures, and how to deal with performance issues.

5
Crazy Bubbles - Your First HTML5 Mobile Game

In the previous chapter, you learned how to implement a Cordova application with the Sencha Touch library. You also learned how to implement a web service and integrate it with the Cordova application. This chapter is about building an HTML5 game and packaging it with PhoneGap to run on mobile devices.

In this chapter, we will cover the following topics:

- Choosing a game framework
- Introduction to Phaser
- What is HTML5 Canvas
- Planning the game
- Generating a Cordova application
- Preparing and creating the game
- Working with assets
- Handling touch events
- Implementing game logic

What game framework to choose

Development of a game from scratch would require a lot of time. It is better to find a framework to accelerate development. There are a lot of ways to build HTML5 games. There are even websites dedicated to list all of the different frameworks available. Here are two such websites:

- http://html5gameengine.com/
- http://html5devstarter.enclavegames.com/

So, which framework should we choose? Some of them are free; some of them are paid. Some of them support WebGL; others work directly with **Document Object Model (DOM)**.

 WebGL is a JavaScript API to work with interactive 3D and 2D computer graphics in a web browser.

I do not really like games using DOM, because it adds limitations to the performance of the application. It is also very hard to create nice animations without glitches. I was not looking for a 3D game framework, but I was interested in good performance. I've spent a lot of time looking into different solutions, and I stopped on Phaser. Why? It's because of the following points:

- It works well on all modern browsers, even on mobile devices.
- It is actively maintained. You can see the increasing number of contributions on GitHub.
- It is powerful. You will see it in the game example we will build.
- It is totally free.
- It has a good documentation.

So, this chapter will be focused on Phaser, a JavaScript canvas and WebGL framework for games. However, look at other JavaScript gaming frameworks as well; there are a lot of them.

What is HTML5 Canvas?

Canvas is the same HTML tag as the others. However, to work with all the possibilities of canvas, we need to use JavaScript. We need to place the `<canvas>` tag on the page in the body section. We then need to access its context with JavaScript and do some drawings on it, almost like painting a real picture.

There is a difference between the canvas element and the canvas context. Context is an object that provides methods to draw on the canvas. Canvas can have two types of context: 2D or WebGL (3D).

Let's look onto one example of the HTML5 Canvas usage:

```
<!DOCTYPE HTML>
<html>
  <head>
    <style>
      body {
        margin: 0px;
        padding: 0px;
      }
    </style>
  </head>
  <body>
    <canvas id="gameCanvas" width="578" height="200"></canvas>
    <script>
      var canvas = document.getElementById('gameCanvas');
      var context = canvas.getContext('2d');
      context.font = '30pt Calibri';
      context.fillStyle = 'red';
      context.fillText('Hello PhoneGap by Example!', 50, 50);
    </script>
  </body>
</html>
```

Here, we put our Canvas tag inside the body, as mentioned earlier. We took this DOM element using the `document.getElementById('gameCanvas')` method. To access Canvas' drawing context, we used the `canvas.getContext('2d')` method. As you can see, we sent the `'2d'` parameter. It specifies that we would like to access Canvas' 2D drawing context. To get access to the WebGL, we can write this `canvas.getContext('3d')` method. Each Canvas element can only have one context. If we use the `getContext()` method multiple times, it would return a reference to the same context object, where:

- `context.font`: Sets the current font properties for context. The font property uses the same syntax as the CSS font property.

- `context.fillStyle`: Sets the color, gradient, or pattern to fill the drawing. In our case, we set it to red.

- `context.fillText`: Draws filled text on canvas. Here, the value first 50 is the x coordinate, where we can start painting the text, and the value second 50 is the y coordinate.

This will result in the following screenshot:

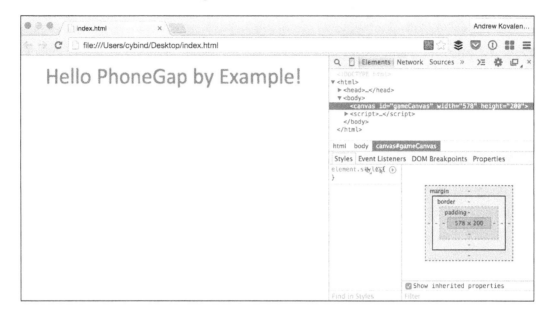

It is just a quick example of what we can do with canvas. There are many more drawings available. We can draw different figures, clean them up, draw pictures on canvas, implement sprite animations, and so on.

Now, let's take a closer look at the Phaser framework and understand how it uses the HTML5 Canvas 2D context for game development needs.

An introduction to Phaser

Phaser was created to build desktop and mobile HTML5 games. It uses Pixi.js (http://www.pixijs.com/) for WebGL and canvas rendering. The framework is growing rapidly. When developing with Phaser, it is possible to use TypeScript or JavaScript. I think it would be a good choice if you are planning to have sounds, collisions, and a lot of different physics in your game. Also, it is good choice for developers who already implemented some games on Flash, because Phaser has a lot of common approaches with Flixel, the game-making library on Flash (http://flixel.org/).

Here are some helpful resources to use as a reference:

- GitHub (https://github.com/photonstorm/phaser)
- Tutorials and documentation (http://phaser.io/learn)
- Examples (http://phaser.io/examples)
- Forum (http://www.html5gamedevs.com/forum/14-phaser/)

The Phaser framework will speed up our development and help with generic tasks to complete the game. Let's plan and develop a simple game, which will cover the basics of using Phaser.

Planning the game

It is always good to have a plan before development. Let's decide what kind of game we would like to implement and describe the main use cases.

I think in the beginning, it should be a simple game. I decided to create a puzzle game, like Gem Match. Here are few examples in the stores:

- https://play.google.com/store/apps/details?id=com.wavelength. gemmatch
- https://itunes.apple.com/WebObjects/MZStore.woa/wa/ viewSoftware?id=719869629&mt=8

The main principle of such games is:

- Swap on adjacent color objects to exchange their positions
- Match three or more colors horizontally or vertically to clear the elements and receive points
- Accumulate as many points as you can

We will implement these principles, but instead of gems, we will use colored bubbles:

Bubbles of the same color will be cleared once they appear in a row or a column. After cleanup, new, randomly generated bubbles will drop from the top. If there are already three or more bubbles of the same color in a row, the player will be able to clear them by a single click or tap.

Generate a Cordova application

Before moving forward, let's prepare a Cordova/PhoneGap application. We can implement the entire game in a web browser, and port it into Cordova in the end. However, I would like to finish this task earlier rather than later.

Let's remember what we did in *Chapter 1, Installing and Configuring PhoneGap*, and run the command to generate the Cordova application:

```
$ cordova create crazy-bubbles com.cybind.crazybubbles CrazyBubbles
```

This command will create a project for us. We now need to add two of our favorite platforms (iOS and Android):

```
$ cd crazy-bubbles
$ cordova platform add ios
$ cordova platform add android
```

To make out status bar look nice, let's just add a status bar plugin, as we did earlier:

```
$ cordova plugin add org.apache.cordova.statusbar
```

Now let's adjust style of the status bar by adding several preferences into config. xml in the root folder:

```
<preference name="StatusBarOverlaysWebView" value="false" />
<preference name="StatusBarBackgroundColor" value="#000000" />
<preference name="StatusBarStyle" value="lightcontent" />
```

Now, we just need to emulate our newly created dummy application:

```
$ cordova emulate ios
```

Finally, we will see a standard UI that we will need to change. Now, we need to start using the Phaser framework.

Getting started with Phaser

It is very easy to get started with Phaser. We can do it in several easy steps.

Download Phaser

Let's download the latest version of the Phaser library from `http://phaser.io/` `download`. Usually, I include the minified version of the `phaser.min.js` library in the HTML file.

Get tools

We need a text editor to edit our HTML/CSS/JavaScript code and a browser with console for debugging. I am using Sublime Text (`http://www.sublimetext.com/`) and Google Chrome.

Use a web server

At this instance, you might have some complicated questions. You might ask, "Do we really need a web server to develop an HTML5 game?", and "Why can't we just develop it in the browser and take all the files needed from the file system?"

The main reason we use a web server is browser security. It is related to the protocol used to access files. When we request anything over the Web, we use HTTP. The server and it's security have enough information to ensure that you can only access the files you are meant to. However, when we open HTML files from our computer's disc, they are loaded via the local filesystem. This filesystem is massively restricted for different reasons. Under `://file`, there are no domains or security levels. So, the code on the page can access a computer's filesystem, and this is not really good for the end user.

As this is dangerous, browsers limit access to the filesystem. However, our game will load different resources: JSON/XML data, images, audio files, and other JavaScript files. To access this data, we need to follow the browser's security practices. This is why we need `http://` access to the game files. Thus, we need a web server.

We can use a lot of different web servers. I am not going to dig deep into the details of the installation and configuration, but here is a list of web servers that could be helpful:

- Local web server packages (all in one):
  - ○ MAMP for Mac (`http://www.mamp.info/`)
  - ○ WAMP for Windows (`http://www.wampserver.com/`)
  - ○ XAMPP for different OS (`https://www.apachefriends.org/`)
- Stand alone web servers:
  - ○ Microsoft IIS (`http://www.iis.net/`)
  - ○ Apache (`http://httpd.apache.org/`)
- A simple HTTP server with Python (if you have Python installed, you can run the `python -m SimpleHTTPServer` command)
- The `http-server` for Node.js (`https://www.npmjs.com/package/http-server`) is a simple, zero-configuration command-line HTTP server
- Simply use the public directory of your Dropbox

Also, it is a good way to run our code in the cloud. There are several JavaScript online-editing tools, such as `JSBin`, `codepen`, and `JSFiddle`. However, these are mainly for quick tests of single scripts. There are several really great cloud IDEs:

- Cloud9 IDE (`https://c9.io/`)
- Codenvy (`https://codenvy.com/`)
- Koding (`https://koding.com/`)

However, I decided to simply use `python -m SimpleHTTPServer`, because I have a Mac, and it has Python installed in it. So, for me, it is really just running the command in the proper folder.

Prepare and create the game

When we executed the Cordova command to generate the application, it already created some HTML, CSS, and JavaScript for us. Let's organize it a little. I would like to see such a structure of the www folder:

```
.
├── assets
│     └── sprites
│            └── spheres.png
├── css
│     └── index.css
├── index.html
└── js
      ├── game.js
      ├── index.js
      └── phaser.js
```

In this structure, we should have the following elements:

- spheres.png: This is our file with different bubble colors
- index.html: This is the main HTML file where we should place the canvas
- index.css: This is our main stylesheet file
- phaser.js: This is the Phaser framework itself
- game.js: This is the file with the game login and all Phaser-related code

Let's modify the index.html file so that the body content looks like this:

```
<div id="phaser"></div>
<script type="text/javascript" src="cordova.js"></script>
<script type="text/javascript" src="js/phaser.js"></script>
<script type="text/javascript" src="js/game.js"></script>
<script type="text/javascript" src="js/index.js"></script>
```

Here, we defined the div element where Phaser will insert the canvas and include all the needed JavaScript files where we described all the logic.

Before moving forward with game logic development let's adjust the `index.js` file so that we can run it in two different ways: on a mobile device and in the desktop's browser. We only need to add several lines of code into the `app.bindEvents` function so that it looks like this:

```
var app = {
    initialize: function() {
        this.bindEvents();
    },
    bindEvents: function() {
        var isBrowser = document.URL.indexOf( 'http://' ) === -1
        && document.URL.indexOf( 'https://' ) === -1;
        if ( isBrowser ) {
            document.addEventListener('deviceready',
            this.onDeviceReady, false);
        } else {
            this.onDeviceReady();
        }
    },
    onDeviceReady: function() {
        app.receivedEvent('deviceready');
    },
    receivedEvent: function(id) {
        initGame();
    }
};
app.initialize();
```

What we added here is just a row, `var isBrowser = document.URL.indexOf( 'http://' ) === -1 && document.URL.indexOf( 'https://' ) === -1`, where we check whether there is `http://` or `https://` in the current URL. In this case, `WebView` wrapped by Cordova will not have the HTTP prefix. Once we have detected that we can run our application in browser we do not create an event listener for the `deviceready` event and simply continue our application execution. We call the `this.onDeviceReady()` function, which calls the `app.receivedEvent('deviceready')` method.

After that, we can go into our `www` directory, which is running under our web server. I decided to use the `SimpleHTTPServer` with Python. So, in my case, it looks like this:

$ python -m SimpleHTTPServer 8000

Now, I am able to see the page with the `http://localhost:8000/` address running.

I am running the page under 8000 port, because on 80 port, I already have another web server running.

In receivedEvent, we are calling the initGame() function from game.js.

The game.js file is prepended with the definitions of global constants and variables. Let's take a closer look at the file and understand what each element means:

```
var game,
    BUBBLE_SIZE = 64,
    BUBBLE_SPACING = 2,
    BUBBLE_SIZE_SPACED = BUBBLE_SIZE + BUBBLE_SPACING,
    BOARD_COLS,
    BOARD_ROWS,
    MATCH_MIN = 3,
    bubbles,
    selectedBubble = null,
    selectedBubbleStartPos = { x: 0, y: 0 },
    tempShiftedBubbleTween,
    allowInput,
    scoreText,
    score = 0;
```

In this file, we have the following elements:

- game: This is a global variable, instance of a Phaser.Game object
- BUBBLE_SIZE: This is the size of the bubble we have in the assets
- BUBBLE_SPACING: This is the spacing between bubbles we want to see on canvas
- BUBBLE_SIZE_SPACED: This is the bubble size with padding
- BOARD_COLS: This is the number of bubble columns on canvas/board
- BOARD_ROWS: This is the number of bubble rows on canvas/board
- MATCH_MIN: This is the minimum number of same color bubbles required in a row/column to be considered a match
- bubbles: This is the Phaser group object in which we will store bubbles
- selectedBubble: This is the global variable where we store selected bubbles to access from different places in the game
- selectedBubbleStartPos: This is the currently selected bubble starting position, which is used to stop the player from moving the bubbles too far

- `tempShiftedBubble`: This is the bubble for replacement. When the player moves the selected bubble, we need to swap the position of the selected bubble with the bubble currently in that position

- `allowInput`: This is used to disable input while bubbles are dropping down and respawning

- `scoreText`: This is the text to display the score on the board

- `score`: This is an actual score value

These values will now be global in our game and help keep everything in sync.

In `initGame()`, we start by initializing Phaser as follows:

```
var width = window.innerWidth - window.innerWidth % BUBBLE_SIZE_
SPACED,
    height = window.innerHeight - window.innerHeight % BUBBLE_SIZE_
SPACED;
game = new Phaser.Game(width, height, Phaser.CANVAS, 'phaser', {
preload: preload, create: create });
```

Where:

- `width`: This is the width of the game.

- `height`: This is the height of the game.

- `Phaser.CANVAS`: This will show to render the game. It could be `Phaser.CANVAS`, `Phaser.WEBGL`, or `Phaser.AUTO`. The recommended parameter is `Phaser.AUTO` which automatically tries to use WebGL. However, if the browser or device doesn't support it, it would fall back on canvas.

- `phaser`: This is the HTML `div` that will contain the game.

- `{ preload: preload, create: create }`: This is our main state.

In the code, we detected the window's inner width and height and calculated the size of canvas we would like to create. It should not be bigger than the browser/device width and height. It should include an integer number of bubbles to place vertically and horizontally.

There are some states in the Phaser game that represent screens. In each state, we can place different logic to run on different scenes. Here are some possible states:

- Game loading screen
- Menu
- Game over screen

There are several main functions we will use when creating the state. However, we will dig only into two of them: `preload()` and `create()`.

`preload()`: Everything in this function is executed in the beginning. This is where we usually load the game's assets (images, sounds, and so on).

`create()`: This function is called after the preload function. Here, we set up the game, display sprites, add labels, and so on.

Now, we need to put some code into the `preload()` and `create()` functions. We will deal with this part in the following section.

Now, we have everything configured properly. We also put together some initial code to start with our real game development. Let's jump into the adventure in the next part of the chapter.

Preloading sprite

First, let's load our `spheres.png` sprite. It is important to load the sprite first to be sure that all pictures are loaded from the server before we start animations. We do not want to see our application without pictures. So, add these lines of code in the `preload()` function:

```
game.load.spritesheet("BUBBLES", "assets/sprites/spheres.png", BUBBLE_
SIZE, BUBBLE_SIZE);
```

Where:

- `BUBBLES`: This is the asset key. We reference to the sprite by this key.
- `assets/sprites/spheres.png`: This is the path to the actual sprite file. It can be vertical, horizontal or a grid.
- `BUBBLE_SIZE`: This is the width and height of the sprite. They are our constants with padding value.

Now, we can refresh the page and see that the file is loaded properly. First, we should go to **Developer Tools | Menu | More Tools | Developer Tools**. We can see it under the **Network** tab:

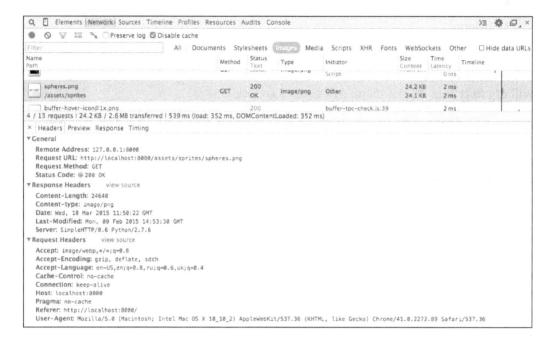

That is all about the `preload()` function.

Displaying sprite

Let's make the content of the `create()` function look like this:

```
spawnBoard();
scoreText = game.add.text(20, game.world.height - 40, 'score: 0', {
font: "20px Arial", fill: "#ffffff", align: "left" });
window.allowInput = true;
```

Where:

- `spawnBoard()`: This fills the screen with as many bubbles as possible.

- `game.add.text`: This adds `score: 0` text with `Arial` font and white color to the bottom-left corner. We will use it later to display the incremented score.

The spawnBoard() method is exactly the function we will use to take out sprite and display it. Let's add the necessary content into this function:

```
BOARD_COLS = Phaser.Math.floor(game.world.width / BUBBLE_SIZE_SPACED);
BOARD_ROWS = Phaser.Math.floor(game.world.height / BUBBLE_SIZE_
SPACED);
bubbles = game.add.group();
for (var i = 0; i < BOARD_COLS; i++)
{
    for (var j = 0; j < BOARD_ROWS; j++)
    {
        var bubble = bubbles.create(i * BUBBLE_SIZE_SPACED, j *
        BUBBLE_SIZE_SPACED, "BUBBLES");
        bubble.name = 'bubble' + i.toString() + 'x' +
        j.toString();
        randomizeBubbleColor(bubble);
        setBubblePos(bubble, i, j);
    }
}
```

In the preceding code, we calculated the number of bubble columns and rows we can put on the stage. We just took the width and height of the canvas and divide it by the bubble size with margin.

After that, we looped through all positions and created the Phaser sprite object. Also, it is saved on top of the bubbles group. The bubble name format I would like to keep is bubble[column number]x[row number] for easier referencing in the future.

Then, we need to assign a random color to the bubble. We can do this using the randomizeBubbleColor function:

```
bubble.frame = game.rnd.integerInRange(0, bubble.animations.frameTotal
  - 1);
```

Here, we take a random frame from all the available sprite frames. To place the bubble on the stage, we just need to assign its posX and posY properties:

```
function setBubblePos(bubble, posX, posY) {
    bubble.posX = posX;
    bubble.posY = posY;
    bubble.id = calcBubbleId(posX, posY);
}
```

In the preceding code, posX and posY are positions on the board and id is an absolute position number, we can then add the following snippet

```
posX + posY * BOARD_COLS
```

Now, we can check the results of our work. If we open `http://localhost:8000/` in the browser, we will be able to see canvas filled with bubbles, as seen in this screenshot:

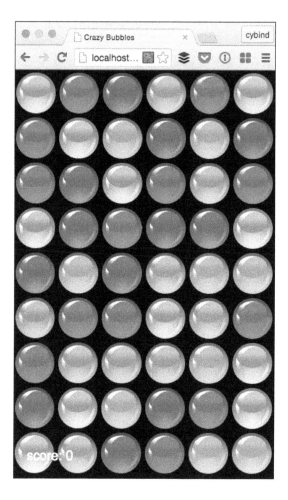

Handling pointer events with Phaser

After we filled the board with bubbles, we need to define user event handlers. We should detect when the user touches the bubble, moves it, and releases it. We should handle both types of events: generated by mouse and by fingers. We have to handle three different events. Let's define three different functions to handle these events:

- `selectBubble`
- `slideBubble`
- `releaseBubble`

First, let's look at the select and release event handlers. We should attach them to each bubble we have on the stage. We can do this when we spawn the board. Let's add the following lines of code to the spawning loop in the `spawnBoard()` function:

```
bubble.inputEnabled = true;
bubble.events.onInputDown.add(selectBubble, this);
bubble.events.onInputUp.add(releaseBubble, this);
```

We enabled bubble for input and attached our event handlers to the `onInputDown` and `onInputUp` events, which are explained here:

- `onInputDown` – This is dispatched when the down event from a pointer is received. The sprite and the pointer object are in the callback.

- `onInputUp` - This is dispatched when the up event from a pointer is received. The only parameter we are handling in the callback is sprite.

Now, let's look at how we can handle the move event on the Canvas. It is different from handling the input down and input up events. We should attach the move handler only once to the entire board. We can do this in the `initGame()` function that is already defined:

```
window.allowInput = true;
game.input.addMoveCallback(slideBubble, this);
```

Where:

- `window.allowInput`: By storing/checking flag in the global variable, we disable input, while bubbles are dropping down and respawning. By default, we enable `window.allowInput`.

- `game.input.addMoveCallback`: This defines `slideBubble` handler for the move event. The pointer, the *x* position of the pointer, the *y* position, and the down state are parameters in the callback.

Now, let's look into the implementation of each event handler. Let's start with the `selectBubble` function:

```
function selectBubble(bubble, pointer) {
    if (window.allowInput)
    {
        window.selectedBubble = bubble;
        window.selectedBubbleStartPos.x = bubble.posX;
        window.selectedBubbleStartPos.y = bubble.posY;
    }
}
```

Here, we simply checked whether we allow the user to pick the bubble and remember the selected bubble and its position into global variables for later use.

The second function is `slideBubble`. This function will be called every time we move the pointer (mouse or finger).

Handling the pointer move event

When we move the pointer, we perform the following tasks:

- Pointer is down
- Select a bubble in the global variable
- Check whether the bubble can be moved into this position
- Swap two neighbor bubbles

The ready `slideBubble` handler looks like this:

```
function slideBubble(pointer, x, y, fromClick) {
    if (window.selectedBubble && pointer.isDown)
    {
        var cursorBubblePosX = getBubblePos(x);
        var cursorBubblePosY = getBubblePos(y);
        if (checkIfBubbleCanBeMovedHere(
            window.selectedBubbleStartPos.x,
            window.selectedBubbleStartPos.y,
            cursorBubblePosX,
            cursorBubblePosY))
        {
            if (cursorBubblePosX !== window.selectedBubble.posX ||
                cursorBubblePosY !== window.selectedBubble.posY)
            {
                bubbles.bringToTop(window.selectedBubble);
                tempShiftedBubble = getBubble(cursorBubblePosX,
                cursorBubblePosY);
                tweenBubblePos(window.selectedBubble,
                cursorBubblePosX, cursorBubblePosY);
                tweenBubblePos(tempShiftedBubble,
                window.selectedBubble.posX, selectedBubble.posY);
                swapBubblePosition(window.selectedBubble,
                tempShiftedBubble);
            }
        }
    }
}
```

However, let's look at each function used here in detail.

Detect the bubble position under the pointer

In the `getBubblePos` function, we convert world coordinates to board position:

```
function getBubblePos(coordinate) {
    return Phaser.Math.floor(coordinate / BUBBLE_SIZE_SPACED);
}
```

We simply divided the x and y coordinates by the bubble size with padding.

Check whether a selected bubble can be moved to a new position

Bubbles can only be moved to one position up/down or left/right, using the following snippet:

```
function checkIfBubbleCanBeMovedHere(fromPosX, fromPosY, toPosX,
toPosY) {
    if (toPosX < 0 ||
        toPosX >= BOARD_COLS ||
        toPosY < 0 ||
        toPosY >= BOARD_ROWS)
    { return false; }
    if (fromPosX === toPosX &&
        fromPosY >= toPosY - 1 &&
        fromPosY <= toPosY + 1)
    { return true; }
    if (fromPosY === toPosY &&
        fromPosX >= toPosX - 1 &&
        fromPosX <= toPosX + 1)
    { return true; }
    return false;
}
```

In the first `if` statement, we checked whether the new position is outside our stage. In the second and third `if` statements, we checked whether we moved only one position left/right or up/down.

If `checkIfBubbleCanBeMovedHere` returns true, we are checking whether the current position is not the position of the selected bubble to avoid the bubble swapping with itself:

```
cursorBubblePosX !== window.selectedBubble.posX || cursorBubblePosY
!== window.selectedBubble.posY
```

Swap bubbles

Now, when all validations passed, we can swap two bubbles.

The `bubbles.bringToTop(window.selectedBubble)` function moves the selected bubble to the top, so that we can see it as it is sliding on top of other bubbles.

When the player moves the selected bubble, we need to swap the position of the selected bubble with the bubble currently in that position. I named this bubble `tempShiftedBubble`:

```
tempShiftedBubble = getBubble(cursorBubblePosX, cursorBubblePosY)
```

With the `getBubble` function, we can find a bubble on the board according to its position on the board:

```
function getBubble(posX, posY) {
    return bubbles.iterate("id", calcBubbleId(posX, posY), Phaser.
Group.RETURN_CHILD);
}
```

Here, we went through the bubbles collection we have, restored ID by *x* and *y* positions, and returned the child bubble by its ID.

The next two rows of the code actually move the animations of the bubbles:

```
tweenBubblePos(window.selectedBubble, cursorBubblePosX,
cursorBubblePosY);
tweenBubblePos(tempShiftedBubble, window.selectedBubble.posX,
selectedBubble.posY);
```

In the first call, we moved the selected bubble to the position of the bubble under the pointer. In the second call, we moved the bubble under the pointer to the position of the selected bubble. To better understand how the `tweenBubblePos` function works, let's look at its content:

```
function tweenBubblePos(bubble, newPosX, newPosY, durationMultiplier)
{
    if (durationMultiplier === null ||
        typeof durationMultiplier === 'undefined')
    {
        durationMultiplier = 1;
    }
    return game.add.tween(bubble).to(
        {
            x: newPosX  * BUBBLE_SIZE_SPACED,
            y: newPosY * BUBBLE_SIZE_SPACED
```

```
        },
        100 * durationMultiplier,
        Phaser.Easing.Linear.None,
        true
    );
}
```

This function has the following parameters:

- `bubble`: This is the element we will animate
- `newPosX`: This will move the element to the x position
- `newPosY`: This will move the element to the y position
- `durationMultiplier`: This is the multiplier we will use to change timeout to start animation

We defaulted `durationMultiplier` to 1 if it is empty. After that, we called the `game.add.tween(bubble).to` single `Phaser` function. A tween allows us to alter one or more properties of a bubble over a defined period of time. This can be used for things such as alpha fading sprites, scaling them, or motion. We pass four parameters into this function:

- The first parameter is an object containing the x and y position properties we want to tween
- The second parameter is the duration of the tween in milliseconds
- The third parameter is the easing function
- The fourth parameter we set to true to allow this tween to start automatically

As a result, our bubble will move from one position to another with linear easing within 100 milliseconds if it is for the first bubble, 200 milliseconds for the second, and so on.

The last function related to sliding is `swapBubblePosition(window.selectedBubble, tempShiftedBubble)`. It exchanges the position values between two bubbles:

```
var tempPosX = bubble1.posX;
var tempPosY = bubble1.posY;
setBubblePos(bubble1, bubble2.posX, bubble2.posY);
setBubblePos(bubble2, tempPosX, tempPosY);
```

The next event that we will trigger is `onInputUp`.

Releasing a bubble

When the mouse or tap is released with a bubble selected, we need to follow these steps:

1. Check for matches

2. Remove matched bubbles

3. Drop down bubbles above the removed bubbles

4. Refill the board

We perform all these steps in the `releaseBubble` function:

```
function releaseBubble(selectedBubble, pointer) {
    checkAndKillBubbleMatches(selectedBubble);
    removeKilledBubbles();
    var dropBubbleDuration = dropBubbles();
    game.time.events.add(dropBubbleDuration * 100, refillBoard);
    window.allowInput = false;
    window.selectedBubble = null;
    window.tempShiftedBubble = null;
}
```

Here, you can see that we refilled the board with some timeout. We did it using the `game.time.events.add` function. The first parameter is timed out for calling the second parameter, that function is to refill the board. We need to be sure that all the existing bubbles have dropped down. In the end of the function, you can see that we disabled input. We enable it only when the board has finished refilling. Let's take a closer look at every item in the `releaseBubble` function.

Check for matches

The first thing we need to do in `checkAndKillBubbleMatches` is count how many bubbles of the same color are above, below, to the left, or the right, using the following code:

```
var countUp = countSameColorBubbles(bubble, 0, -1);
var countDown = countSameColorBubbles(bubble, 0, 1);
var countLeft = countSameColorBubbles(bubble, -1, 0);
var countRight = countSameColorBubbles(bubble, 1, 0);
var countHoriz = countLeft + countRight + 1;
var countVert = countUp + countDown + 1;
```

Where `countSameColorBubbles` is:

```
function countSameColorBubbles(startBubble, moveX, moveY) {

    var curX = startBubble.posX + moveX;
    var curY = startBubble.posY + moveY;
    var count = 0;

    while (curX >= 0 &&
           curY >= 0 &&
           curX < BOARD_COLS &&
           curY < BOARD_ROWS &&
           getBubbleColor(getBubble(curX, curY)) ===
           getBubbleColor(startBubble))
    {
        count++;
        curX += moveX;
        curY += moveY;
    }

    return count;
}
```

Here, we counted how many bubbles of the same color lie in a given direction. There are four different directions. Hence, we called this function four times. If `moveX = 0` and `moveY = -1`, it will count how many bubbles of the same color lie to the bottom of the bubble. We stop counting as soon as a bubble of a different color or the stage end is encountered.

If more than three bubbles match horizontally or vertically, kill them.

```
if (countVert >= MATCH_MIN)
{
    killBubbleRange(
      bubble.posX,
      bubble.posY - countUp,
      bubble.posX,
      bubble.posY + countDown
    );
}
if (countHoriz >= MATCH_MIN)
{
    killBubbleRange(
      bubble.posX - countLeft,
      bubble.posY,
```

```
        bubble.posX + countRight,
        bubble.posY
    );
}
```

Where `killBubbleRange` is:

```
function killBubbleRange(fromX, fromY, toX, toY) {
    fromX = Phaser.Math.clamp(fromX, 0, BOARD_COLS - 1);
    fromY = Phaser.Math.clamp(fromY , 0, BOARD_ROWS - 1);
    toX = Phaser.Math.clamp(toX, 0, BOARD_COLS - 1);
    toY = Phaser.Math.clamp(toY, 0, BOARD_ROWS - 1);

    for (var i = fromX; i <= toX; i++)
    {
        for (var j = fromY; j <= toY; j++)
        {
            var bubble = getBubble(i, j);
            bubble.kill();
        }
    }
}
```

Where:

- `fromX, fromY`: This is the starting position from which we should kill bubbles

- `toX, toY`: This is the end position from which we should kill bubbles

We looped through these positions, retrieving bubbles by this position and removing them by calling the `.kill()` method. It will set the `alive` property of the bubble to `false`.

If no match was made, move the bubbles back into their starting positions:

```
if (countVert < MATCH_MIN && countHoriz < MATCH_MIN)
{
    if (bubble.posX !== window.selectedBubbleStartPos.x ||
        bubble.posY !== window.selectedBubbleStartPos.y)
    {
        tweenBubblePos(
            bubble,
            window.selectedBubbleStartPos.x,
            window.selectedBubbleStartPos.y
        );
```

```
        if (tempShiftedBubble !== null)
        {
            tweenBubblePos(
                tempShiftedBubble,
                bubble.posX,
                bubble.posY
            );
        }
        swapBubblePosition(bubble, tempShiftedBubble);
    }
}
```

You can see as we tween back for both `bubble` and `tempShiftedBubble` and swapping position values back as well.

Remove matched bubbles

We called the `kill()` method for each bubble that we needed to remove. Now, we need to hide these bubbles on the board. We can do this by simply moving them away from the board to the `-1` column and `-1` row:

```
function removeKilledBubbles() {
    bubbles.forEach(function(bubble) {
        if (!bubble.alive) {
            setBubblePos(bubble, -1,-1);
        }
    });
}
```

Drop down bubbles above the removed bubbles

Now, we will look for bubbles with empty space underneath them and move them down, using the following code:

```
function dropBubbles() {
    var dropRowCountMax = 0;
    for (var i = 0; i < BOARD_COLS; i++)
    {
        var dropRowCount = 0;
        for (var j = BOARD_ROWS - 1; j >= 0; j--)
        {
            var bubble = getBubble(i, j);
            if (bubble === null)
```

```
        {
            dropRowCount++;
        }
        else if (dropRowCount > 0)
        {
            setBubblePos(
                bubble,
                bubble.posX,
                bubble.posY + dropRowCount
            );
            tweenBubblePos(
                bubble,
                bubble.posX,
                bubble.posY,
                dropRowCount
            );
        }
    }
    dropRowCountMax = Math.max(dropRowCount, dropRowCountMax);
}

return dropRowCountMax;
}
```

Here, we looped through all cells and tried to get the bubble from its position. If the bubble is null, then we would increase `dropRowCount`. Once we reached the existing bubble in a column, we set the bubble position and moved it down to fill empty space.

We calculate `dropRowCount` for another reason as well. On the next stage, when we will fill the board, we need to wait until all the existing bubbles have dropped down. Remember that animation for each bubble takes 100 milliseconds, that is defined in the following code:

```
game.time.events.add(dropBubbleDuration * 100, refillBoard);
```

Refill the board

In the `refillBoard` function, we look for any empty spots on the board and spawn new bubbles in their place that fall down from above, using the following code:

```
function refillBoard() {

    var maxBubblesMissingFromCol = 0;
```

```
for (var i = 0; i < BOARD_COLS; i++)
{
    var bubblesMissingFromCol = 0;

    for (var j = BOARD_ROWS - 1; j >= 0; j--)
    {
        var bubble = getBubble(i, j);

        if (bubble === null)
        {
            bubblesMissingFromCol++;
            bubble = bubbles.getFirstDead();
            bubble.reset(
                i * BUBBLE_SIZE_SPACED,
                -bubblesMissingFromCol * BUBBLE_SIZE_SPACED
            );
            randomizeBubbleColor(bubble);
            setBubblePos(bubble, i, j);
            tweenBubblePos(
                bubble,
                bubble.posX,
                bubble.posY,
                bubblesMissingFromCol * 2
            );
        }
    }

    maxBubblesMissingFromCol =
    Math.max(maxBubblesMissingFromCol, bubblesMissingFromCol);
}

game.time.events.add(maxBubblesMissingFromCol * 2 * 100,
boardRefilled);
}
```

Again, we are looping through the board. Once we find bubble === null parameter, we increment bubblesMissingFromCol for later calculation of the timeout to call the complete refill action. To better utilize memory, we do not create a new sprite. Instead, we took the sprite we first killed earlier, reset it on the needed position, and randomized its color. Eventually, we called tween, which gracefully slides in new bubbles from the top one by one.

Once all animations are done, we called the `boardRefilled` function, where we re-enable player input:

```
window.allowInput = true;
```

Calculate score

In this section, you will learn how to add a function that increases the score when a player removes bubbles. We already have the `removeKilledBubbles()` function, which removes killed bubbles from the stage. So, we can add our score calculation and display logic here as well.

```
bubbles.forEach(function(bubble) {
    if (!bubble.alive) {
        setBubblePos(bubble, -1,-1);
        score++;
    }
});
scoreText.text = 'score: ' + score;
```

Here, you can see that we incremented the `score` global variable and updated the text of the `scoreText` object. Now, you can refresh the browser, and as you remove each bubble, you'll see the score increase by 1.

Running the application on the mobile

The most exciting moment of game development is when you have finished creating your game and are ready to enjoy playing it. Let's run the game on our plugged iPhone, using the following command:

```
$ cordova run ios
```

Voila! We see our game on the iPhone and can even play and score some points:

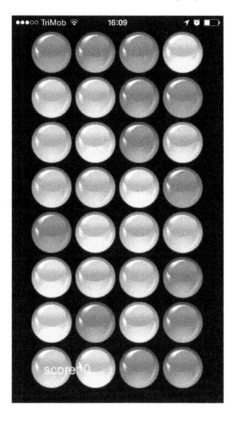

Summary

At this point, the game is basically created, and hopefully, you've learned how some of the core mechanics of Phaser work.

In this chapter, we covered important assets, and you learned how to create sprites, create tweens, use timers, and so on. You also learned how to wrap your game with Cordova/PhoneGap, for distribution as a native mobile application.

In the next chapter, you will learn how to integrate your game with third-party services, such as Facebook and Twitter.

6
Share Your Crazy Bubbles Game Result on Social Networks

In the previous chapter, you learned how to create a HTML5 game called Crazy Bubbles, where the main idea is to collect as many bubbles of the same color as possible. In this chapter, we will communicate with several of the most popular social networks: Twitter, Facebook, and Instagram. We will implement a game over criteria for the game and review several options to share our results on social networks. Here is a short list of the topics we will cover in this chapter:

- Checking the **game over** criteria
- Posting a new message on Twitter and Facebook
- Communicating with the Instagram application

Implementing the game over screen

Before sharing the game results, we should implement the game over scenario. In our case, it is probably an expensive operation. We should check all the possible moves on the stage. If there are no moves a user can make to clear the next set of color bubbles, we should notify the user that the game is finished and show the share dialog box to him. So, let's develop an algorithm to check the possible moves.

I will separate the logic into two different scenarios:

- When we check vertical identity
- When we check horizontal identity

For both scenarios, the logic is pretty much the same. Just the x and y coordinates should change in the calculations.

The vertical scenario

In the following image, you can see all the possible positions of bubbles with the same color:

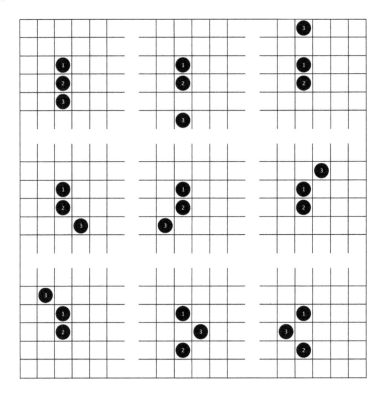

Here, you can see nine different possible combinations. I displayed only three same color bubbles in each combination, because it is enough to detect the possible successful move.

Let's look at combinations from left to the right and from top to the bottom.

The first seven combinations are based on validation if the two bubbles already are neighbors in a column. We are just checking whether the third bubble is:

- Right next to the second bubble
- In the second cell underneath the second bubble
- In the second cell above the first bubble
- One cell down and one cell right to the second bubble
- One cell down and one cell left to the second bubble
- One cell up and one cell right before the first bubble
- One cell up and one cell left before the first bubble

However, if two bubbles are not together in the column, then we could check for a same color bubble in the second cell underneath the first bubble. If we find the needed color there, then we have only two possible combinations where we can look for the third bubble with the same color:

- One cell down and one cell right to the first bubble
- One cell down and one cell left to the first bubble

The horizontal scenario

In the horizontal scenario, we will use a similar approach. Let's look at the following image:

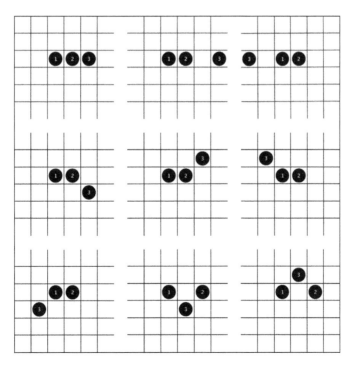

In the first seven combinations, we saw all the possible positions to check whether the first two bubbles are close. We also checked whether the third bubble is:

- Right next to the second bubble
- In the second cell to the right of the second bubble
- In the second cell before the first bubble
- One cell down and one cell right to the second bubble
- One cell up and one cell right to the second bubble
- One cell up and one cell left before the first bubble
- One cell down and one cell left before the first bubble

If two bubbles are not close together, we could check whether the third bubble is:

- One cell down and one cell right to the first bubble
- One cell up and one cell right to the first bubble

So, we have explored all the possible successful combinations. Now, we need to code this logic.

Coding the logic

First of all, we need to create a text with the words Game Over. We should place it on the screen as soon as we find at least one suitable successful combination. We could do this using the create() function:

```
function create() {
    // ...
    gameOverText = game.add.text(game.world.centerX, 100, 'Game
    Over!', { font: "40px Arial", fill: "#ffffff", align: "center" });
    gameOverText.visible = false;
    gameOverText.anchor.setTo(0.5, 0.5);
    // ...
}
```

Here, we created the Phaser text object with the text Game Over! placed horizontally in the center and 100 pixels from the top. The last parameter in the game.add.text function is the style object containing style attributes such as font, font size, and so on. After that, we hid the text object so that it is not visible while we are playing the game. The last row is the gameOverText.anchor.setTo(0.5, 0.5) function call where we set the object's center offset to the center. In this case, we can easily center the object on the canvas.

We need another help function to get the bubble color by its position, as shown in the following code:

```
function getBubbleColorByPos(x, y) {
    var color = null;
    if (x >= 0 && y >= 0 && x < BOARD_COLS && y < BOARD_ROWS) {
        var bubble = getBubble(x, y);
        color = getBubbleColor(bubble);
    }
    return color;
}
```

In the function, we checked whether the x or y position is not out of the stage. We simply used two earlier defined functions getBubble and getBubbleColor to get the bubble by its position and detect its color.

Now, we need to check all the possible combinations for vertical and horizontal scenarios.

```
function checkPossibleMoves() {
    if (loopAndCheckCombinations('v')) return true;
    if (loopAndCheckCombinations('h')) return true;
    return false;
}
```

The checkPossibleMoves() method returns false if any of the inner steps return true.

In the loopAndCheckCombinations function, we passed only one parameter with value v or h, where v means vertical, and h means horizontal. The logic for the two directions is pretty much the same. Only the x and y coordinates should be switched.

Let's look at the loopAndCheckCombinations function in detail:

```
function loopAndCheckCombinations(direction) {

    var verticalCheckCells1 = [
        {i:0,  j:3 },
        {i:0,  j:-2},
        {i:1,  j:2 },
        {i:-1, j:2 },
        {i:1,  j:-1},
        {i:-1, j:-1}
    ];
```

```
    var verticalCheckCells2 = [
        {i:1,  j:1 },
        {i:-1, j:1 }
    ];

    for (var i = 0; i < BOARD_COLS; i++)
    {
        for (var j = 0; j < BOARD_ROWS; j++)
        {
            // ...
        }
    }

    return false;
}
```

Here, we defined two arrays: `verticalCheckCells1` and `verticalCheckCells2`.
The first array contains objects with coordinates to check whether, around the
current position, there are two same color bubbles in neighboring positions. The
second array contains only two objects with coordinates to check whether, within the
current position, there are two bubbles of the same color coming out of one cell. It is
exactly the same position we saw in the figures earlier in the chapter.

Once we define these arrays, we loop through each bubble on the stage and run the
following lines of code:

```
var x1, x2, y1, y2;
if (direction === 'v') {
    x1 = x2 = i;
    y1 = j+1;
    y2 = j+2;
} else if (direction === 'h') {
    x1 = i+1;
    x2 = i+2;
    y1 = y2 = j;
}

var color0 = getBubbleColorByPos(i, j);
var color1 = getBubbleColorByPos(x1, y1);
var color2 = getBubbleColorByPos(x2, y2);

if (color0 == color1 && color0 == color2) {
    return true;
} else if (color0 == color1) {
```

```
    if (checkCombinations(i, j, 0, color0, verticalCheckCells1,
    direction)) return true;
} else if (color0 == color2) {
    if (checkCombinations(i, j, 0, color0, verticalCheckCells2,
    direction)) return true;
}
```

Here, we checked the mode in which we should work: vertical or horizontal. If we are in the vertical mode, we take three bubble colors from the current position down. If we are in the horizontal mode, we take three bubble colors from the current position right. After that, we compare colors. If there are three same color bubbles in a row or column, we immediately return true, and all our logic work is finished. If two same color bubbles are neighbors, we continue logic execution against the verticalCheckCells1 array. If there are two same color bubbles in one position, we continue checking against the verticalCheckCells2 array.

The checkCombinations function is a recursive one. It loops through the verticalCheckCells1 or verticalCheckCells2 array in context of the current bubble. Let's take a closer look at this function:

```
function checkCombinations(i, j, k, color, pos, direction) {
    if (k < pos.length) {
        var x, y;
        if (direction === 'v') {
            x = pos[k].i;
            y = pos[k].j;
        } else if (direction === 'h') {
            x = pos[k].j;
            y = pos[k].i;
        }
        var color2 = getBubbleColorByPos(i+x, j+y);
        if (color == color2) {
            return true;
        } else {
            return checkCombinations(i, j, k+1, color, pos,
            direction);
        }
    } else {
        return false;
    }
}
```

Here:

- `i`: Here the x coordinate is for vertical direction and y is for horizontal
- `j`: Here the y coordinate is for vertical direction and x is for horizontal
- `k`: This is the index of the object in the `pos` array
- `color`: This is the color of the bubble to compare with
- `pos`: This is the array with object coordinates to check
- `direction`: This can be vertical or horizontal

In the beginning, we checked whether `k < pos.length` and continue execution if it is true. If we have not reached the end of the combination checking array, then we detect the current direction and swap the x and y coordinates of the objects if we are in the horizontal mode. After that, we will detect the color of the bubble in the position from the checking array. If we find the same color bubble, we return true; else, we continue recursion.

We continue to loop through all positions on the stage until we find a successful combination or reach the end of the board. It means that there is no successful combination once we reach the last bubble on the board.

Once the `checkPossibleMoves()` function returns false, we can show the game over screen. I decided to put the logic in the `boardRefilled()` function. This function is called exactly when we need to check combinations. So, let's see how the function looks now:

```
function boardRefilled() {
    if (checkPossibleMoves()) {
        window.allowInput = true;
    } else {
        console.log('GAME OVER!!!');
        gameOver();
    }
}
```

In this function, we allow user input if there are any possible moves or call the `gameOver()` function to show the game over screen, as shown in the following code:

```
function gameOver() {
    gameOverText.visible = true;
    window.allowInput = false;
    document.getElementById('game-over').style.display = "block";
}
```

Here we display the text object, disable user input, and display the `div` tag with three buttons.

It is a simple layout with the following content:

```
<div id="game-over">
    <div class="wrapper">
        <a href="#" id="restart"><img src="assets/restart.png"
        width="30"></a>
        <a href="#" id="share"><img src="assets/share.png"
        width="30"></a>
        <a href="#" id="share-instagram"><img
        src="assets/instagram-icon.png" width="30"></a>
    </div>
</div>
```

Here we defined three links that will help user to restart the game, share to different social networks, and to pass the picture to the installed Instagram application.

The game over screen looks similar to this screenshot:

Implementing game restart

We should click on the restart icon on the game over screen. It is the first button in the row below **GAME OVER!**, we will implement the `restart` function with the following steps:

1. Let's add the `restart` function as an event listener for the click event:

   ```
   document.getElementById('restart').addEventListener("click",
   restart);
   ```

   The `restart` function looks like this:

   ```
   function restart() {
       gameOverText.visible = false;
       document.getElementById('game-over').style.display =
       "none";
       killBubbleRange(0, 0, BOARD_COLS - 1, BOARD_ROWS - 1);
       removeKilledBubbles();
       score = 0;
       scoreText.text = 'score: ' + score;
       refillBoard();
       return false;
   }
   ```

2. In the function, we hide the **Game Over!** text as well as the following three icons:

   - Restart
   - Share
   - Share with Instagram

3. After that, we kill all the bubbles on the board and remove them from the stage using the `killBubbleRange` and `removeKilledBubbles` functions, which are already defined. Also, we reset the score value and score text on the screen.

4. Finally, we call the function to refill the board.

When you click on the restart button, you will see a clean board and bubbles will start dropping down from the top. For example, here is a single frame from the restart process:

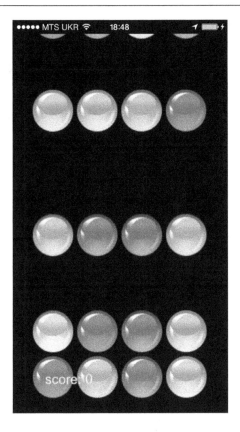

Sharing on Twitter, Facebook, and other social media

It is always good to add some social media to the PhoneGap/Cordova application. We can draw attention to our application when users do the promotional work for us. What we need is to add share buttons for Facebook, Twitter, Tumblr, and other social media on our application. But how do we make it work for PhoneGap? We can use the PhoneGap Social Sharing plugin.

 You can read details about this plugin at `https://github.com/EddyVerbruggen/SocialSharing-PhoneGap-Plugin`.

We get the **Game Over!** screen implemented so that we can use this plugin now to provide the ability to share our achieved results.

First of all, let's install the plugin. We can do this using the well-known method, which is as follows:

```
$ cordova plugin add nl.x-services.plugins.socialsharing
```

With this plugin, we can have our iOS, Android, or Windows Phone open the native share widget. The user can share results on the social media of his choice.

With the plugin, we can share:

- Text
- Text with subject
- Link
- Image or other file (`.pdf`, `.txt`, and so on)

> It is possible to share files from the Internet, the local filesystem, or from the www folder.

Before moving forward with sharing, we have to attach an event handler or click first and grab the picture of the game screen.

We can grab the content of the HTML5 Canvas with the `toDataURL()` function. The result of the `toDataURL()` function is a 64-bit encoded PNG URL.

We can grab image data URL in the JPEG format as well. We can pass `image/jpeg` as the first argument in the `toDataURL()` method. In addition, we can control the quality of the JPEG image. To do this, we just need to pass either `0` or `1` as the second argument.

Here is our full code to handle sharing on social media:

```
document.getElementById('share').addEventListener("click", function()
{
    var img = game.canvas.toDataURL();
    window.plugins.socialsharing.share('I just got ' + score + '
    points in Crazy Bubbles!', 'Crazy Bubbles', img,
    'http://www.crazy-bubbles.com/');
    return false;
});
```

As you can see here, we took the 64-bit encoded PNG URL of the canvas and sent it along with other information in the `window.plugins.socialsharing.share` call. The sharing function has the following arguments in it:

- **Message**: `'I just got ' + score + ' points in Crazy Bubbles!'`
- **Subject**: `'Crazy Bubbles'`
- **Image**: `img`
- **Link**: `'http://www.crazy-bubbles.com/'`

Once the user clicks on the share button, they will see something like following screen in iOS:

The share options can vary. They depend on what is available on the device applications and what has been set up in the device settings. Also, you may realize that not all fields pass to sharing screens. There are different sets of arguments that can be passed to different applications. Here are some examples:

- **Mail**: This provide message, subject, file
- **Twitter**: This contains message, image, link
- **Facebook iOS**: This provides message, image, link
- **Facebook Android**: This provides a link or image only

The sharing screen for Twitter and Facebook in iOS might look like this:

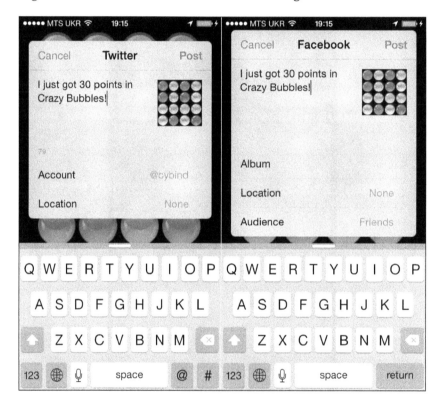

Sharing on Instagram

Another useful plugin for sharing is the Instagram plugin for Cordova.

 You can read more about this plugin at https://github.com/vstirbu/InstagramPlugin.

It supports both iOS and Android platforms.

The requirement for the plugin is installed in the Instagram application. It will show a notification if Instagram is not installed.

Let's install the plugin as we usually do it:

```
$ cordova plugins add https://github.com/vstirbu/InstagramPlugin
```

We already know how to add an event handler on the Instagram share button. Take a Base64 data URL image of the canvas and send it along with the caption:

```
document.getElementById('share-instagram').addEventListener("click",
function() {
    var img = game.canvas.toDataURL();
    var caption = 'I just got ' + score + ' points in Crazy
    Bubbles!';
    Instagram.share(img, caption, function (err) {
        if (err) {
            console.log("not shared");
        } else {
            console.log("shared");
        }
    });
    return false;
});
```

Here, we called `Instagram.share` with the following arguments:

- **Image**: `img`
- **Caption**: `'I just got ' + score + ' points in Crazy Bubbles!'`
- **Callback**: Obtaining the `err` variable if there is a sharing error

Once the user clicks on the share to Instagram button, they will be able to see the following screen:

It is pretty similar to the previous share screen, except it shows only one available option: Instagram.

When a user clicks on the Instagram icon, it opens the Instagram application and brings the picture and caption to it:

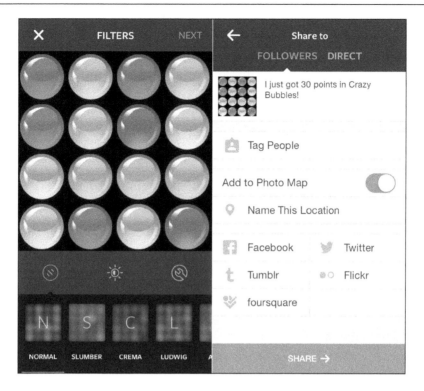

That is it! We can successfully apply all Instagram features to the picture and share it.

Summary

In this chapter, we finished our game and implemented the Game Over logic. We successfully created a fully working application that shares messages on Twitter, Facebook, and other social media. Essential work was done by external modules, which again proves that the PhoneGap/Cordova ecosystem is really flexible and provides everything you need to develop top-notch hybrid mobile applications.

In the next chapter, you will find out how to use the WebRTC API along with PhoneGap to build a video/audio communication application.

7
Building a Real-time Communication Application – Pumpidu

In the previous chapters, you learned how to use Sencha Touch with the Cordova/PhoneGap application, how to use PhoneGap/Cordova plugins, and how to develop HTML5 mobile games with Phaser and wrap them with Cordova. Also, you learned how to integrate your mobile application with the custom service built with Node.js and how to add some social media functions to it. It is clear that PhoneGap/Cordova, as a platform, becomes more powerful when we use it with other web technologies.

In this chapter, we will revisit Node.js to implement the server side part for our application. Also, in this chapter, we will go through the process of developing video real-time communication / audio real-time communication web and mobile applications. For this, we will use the modern web browser's API named WebRTC. At the end of this chapter, we want to achieve a mobile application that calls another mobile or web client and establishes a video/audio communication channel.

In this chapter, we will cover the following topics:

- WebRTC fundamentals
- Client-side implementation
- Server-side implementation
- Using PeerJS to simplify WebRTC development
- Reviewing other tools to build WebRTC-based applications

I assigned the name `Pumpidu` to the project. It is nothing special, just a fictional name. I will make this application available at `http://pumpidu.com`.

WebRTC fundamentals

WebRTC is one of the most interesting inventions in the Web lately.

WebRTC (real-time communication) is an Internet protocol. It is an open source software designed to organize streaming data between browsers and other peer-to-peer connected applications.

Its inclusion in the W3C recommendation supports Google, Mozilla, and Opera. After the integration of WebRTC into Chrome (and possibly also in a number of other popular browsers), Google can compete with Skype. Third-party web developers can create their voice-/video-communication applications based on the technology WebRTC.

WebRTC is a collection of standards, protocols, and JavaScript API. This combination provides peer-to-peer audio, video, and data transfer between peers. WebRTC simplifies the work with the abstraction of JavaScript API.

It's not just client-server connection; it is a bit more complicated.

The main task of WebRTC is not only real-time communication between browsers. WebRTC was created with the idea of integrating it with existing communication systems such as VoIP, SIP clients, and regular phones.

WebRTC consists of three major components:

- `MediaStream`: This is used to get audio and video streams
- `RTCPeerConnection`: With this we can perform audio and video data transmission
- `RTCDataChannel`: With this we can perform custom data format transfer

The following diagram depicts the preceding points:

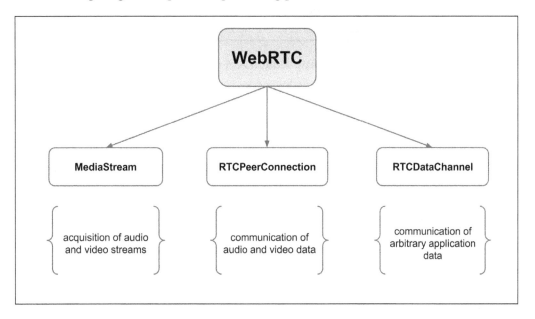

WebRTC audio and video engines

Implementation of a conference call in the browser requires access to capture both audio and video. Audio and video must be processed to improve the quality. Synchronized and outgoing bitrate must adapt to the bandwidth and latency capacity channels.

On the side of the recipient, it is vice versa. It is necessary to decode the stream in real time and adopt the instability of the network. It is a difficult task, and WebRTC solves it well.

WebRTC uses two audio codecs created in GIPS, and the video format VP8 (WebM).

WebRTC captures video, removes noise and improves image, removes echo and synchronizes with the video, and processes errors in network instability:

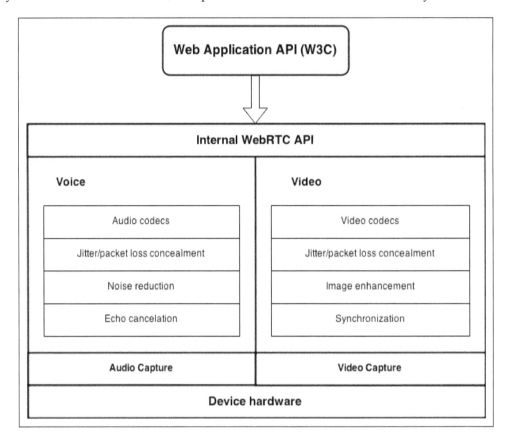

The WebRTC protocol stack

UDP is just for WebRTC, but it also needs to implement a lot of things to ensure that they are there in the UDP. We need a big package of protocols and services for support. The package includes the following elements:

- **Interactive Connectivity Establishment (ICE)**
    - ◦ **Session Traversal Utilities for NAT (STUN)**
    - ◦ **Traversal Using Relays around NAT (TURN)**
- **Datagram Transport Layer Security (DTLS)**

- **Stream Control Transport Protocol (SCTP)**
- **Secure Real-time Transport Protocol (SRTP)**

ICE, STUN, and TURN are needed to create and manage peer-to-peer connections over UDP. DTLS is used to protect data; it is a requirement for WebRTC. SCTP and SRTP are software protocols for the implementation of the multiplexing of the various streams. They provide congestion and flow control, and the possibility of a partial delivery of data.

			RTCPeerConnection	DataChannel
XHR	SSE	WebSocket	SRTP	SCTP
HTTP			Session (**DTLS**) - mandatory	
Session (TLS) - optional			ICE, STUN, TURN	
Transport (TCP)			Transport (**UDP**)	
Network (IP)				

The RTCPeerConnection API

It is pretty easy to establish peer-to-peer connection with the WebRTC API.

The ICE agent automatically starts the search process of the IP and the port of possible candidates. The following actions are involved in doing so:

- The ICE agent asks the operating system about the local IP.
- The ICE agent requests a remote STUN server to get a public IP and port of another point.
- If configured, ICE agents apply the TURN server as an extreme candidate. If a peer-to-peer connection is not established, the connection would be established through an alternative channel.

The following diagram depicts the preceding points:

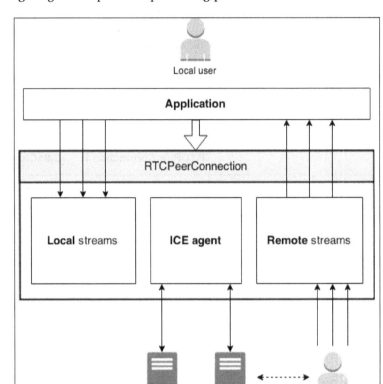

The WebRTC browser support scorecard

WebRTC is not supported by some of the browsers yet. We can check the current state of it at `http://iswebrtcreadyyet.com/`.

Mainly, Chrome, Firefox, and Opera support WebRTC. We will discuss why we need to understand what browsers support WebRTC in the following section.

What is Crosswalk and why we need it?

When we develop with PhoneGap/Cordova, we use the standard WebView that the system provides for us. The situation is not bad with iOS. WebView in iOS from version to version works almost the same. There are only small changes and improvements provided for now.

However, with Android, the situation is totally different. In the different variations of the platform, different versions of WebView are used. There are differences in:

- JavaScript API
- CSS properties support and syntax
- Specific interface that renders flow

Many other differences and odds can also be found.

This is where Crosswalk is very handy. Like WebView, it uses the latest version of Google Chromium. Crosswalk adds the following benefits:

- Same WebView on all Android 4.x platforms
- Available Chrome DevTools
- Great performance of JavaScript, HTML, and CSS

It is like running a Google Chrome instance in your native mobile application.

Here are several other benefits helpful for us:

HTML5 feature	Without the Crosswalk project	With the Crosswalk project
WebRTC	-	+
WebGL	-	+
Vibration API	-	+
Presentation API	-	+
WebView updates	-	+

What is critical for us is the support of WebRTC. In default WebView that PhoneGap/Cordova uses for Android, there is not support of WebRTC. So, we will now add Crosswalk to the Cordova application.

Adding Crosswalk support to the Cordova application

Let's start creating our application. We need to create a client folder and initialize our Cordova application here:

```
$ cordova create client com.cybind.pumpidu Pumpidu
```

It will create the following folder structure for us:

```
.
├── config.xml
├── hooks
├── platforms
├── plugins
└── www
```

Now, let's add an Android platform:

```
$ cordova platform add android
```

In Cordova, for Android version after 4.0.0, there are a number of important changes. The most important one for us is that it adds first-class support for Crosswalk. With the support of pluggable WebViews, we can easily add Crosswalk as follows:

```
$ cordova plugin add https://github.com/MobileChromeApps/cordova-plugin-crosswalk-webview.git#1.0.0
```

It is a great idea to have an option to select different WebViews simply by installing plugins.

Building our first real-time communication application

We have successfully created a basic Cordova application with Crosswalk support. Before moving forward, let's create a simple server to handle our needs for the application we are building.

Server side

As we discussed earlier, we need to build a signaling (STUN) server. We can do this easily with Node.js. For now, we are not digging into the TURN server setup, because we will test it in the local network without enterprise NAT traversal and firewalls.

We need create a simple Node.js application with socket.io. Let's start by creating package.json. We can do this easily with the npm init command. Follow several simple steps, and we will have something similar to these lines of code:

```
{
  "name": "pumpidu",
  "version": "1.0.0",
```

```
  "description": "WebRTC project",
  "main": "server.js",
  "dependencies": {},
  "devDependencies": {},
  "scripts": {
    "test": "echo \"Error: no test specified\" && exit 1",
    "start": "node server.js"
  },
  "repository": {
    "type": "git",
    "url": "https://github.com/cybind/pumpidu.git"
  },
  "keywords": [
    "WebRTC"
  ],
  "author": "Andrew Kovalenko",
  "license": "ISC",
  "bugs": {
    "url": "https://github.com/cybind/pumpidu/issues"
  },
  "homepage": "https://github.com/cybind/pumpidu"
}
```

You can see some generic information about the project here. It is the application name, version, some description, main file, production and development dependencies, script to run the test and start applications, links to the `git` repository, `author`, and so on.

Let's add `socket.io` to our project and to the dependencies:

```
$ npm install socket.io --save
```

After that, we will create our single file in the project: `server.js`. For now, it will be empty. So, the structure of the server application folder looks like this:

```
.
├── node_modules
├── package.json
└── server.js
```

Now, we need to add some working script inside the folder. We can use `socket.io` with the Node.js `http` server. Let's use the `require` command, and start our `http` server:

```
var http = require('http');
var SERVER_PORT = '1234';
var app = http.createServer().listen(SERVER_PORT);
```

Now, we can use the `require` command on `socket.io` and add a listener to our `http` application:

```
var io = require('socket.io').listen(app);
```

Next, we will attach an event handler on the `message` event once the `socket` connection is established:

```
io.sockets.on('connection', function(socket) {

    socket.on('message', function(message) {
        socket.broadcast.emit('message', message);
    });

});
```

You can see here that once we got the `message` notification, we generated a broadcast notification with the same content we got with message. It will send a message to everyone else, except the socket that starts it.

And that is it! This is the minimal server setup we could prepare for our STUN server. Now, we just need to start it using the following command:

```
$ node server.js
```

Let's continue and build the client side of our WebRTC application.

Client side

We already have generated the Cordova/PhoneGap application. So, we need to add WebRTC components to it.

First of all, we need to add video elements to the page:

```
<video id="localVideo" autoplay muted></video>
<video id="remoteVideo" autoplay></video>
<button id="callButton">Call</button>
```

Where:

- `localVideo`: This is the HTML5 video tag where we will display our local video
- `remoteVideo`: This is the HTML5 video tag where we will display the video and play sound from another peer
- `callButton`: This is a button responsible for initiating a call with another peer

Now, we will add the script elements to the page:

```
<script type="text/javascript" src="cordova.js"></script>
<script type="text/javascript" src="js/socket.io.js"></script>
<script type="text/javascript" src="js/index.js"></script>
```

The `cordova.js` file is a standard library added by Cordova CLI in the application-generation stage. We can download the `Socket.io` client library from `http://socket.io/download/`. There are several options to serve it from CDN, but we would like our PhoneGap/Cordova application to have as few external dependencies as possible. So, I just downloaded the file and placed it in the project. The `index.js` file is our main file where all the client-side logic is placed.

Before looking into the JavaScript logic, let's define some CSS styles for our application first. We will add them in the `css/index.css` file and include it in `index.html` with the following tag:

```
<link rel="stylesheet" type="text/css" href="css/index.css">
```

In the following style, we will make `remoteVideo` of full width:

```
#remoteVideo
{
    width: 100%;
    height: 100vh;
    background: #000;
}
```

Make a local video of `20%` width and place it in the bottom-right corner:

```
#localVideo
{
    position: absolute;
    right: 1.1em;
    bottom: 1em;
    width: 20%;
    border: 1px solid #333;
    background: #000;
}
```

For both local and remote video elements, we will add a black background, so it is nice when there is no video call established.

The `callButton` tag will be with a green background and rounded buttons. We will place it in the center at the bottom of the screen.

```
#callButton
{
    font-size: 1.5em;
    position: absolute;
    bottom: 5%;
    left: 50%;
    display: none;
    width: 4em;
    height: 2em;
    margin-left: -2em;
    border: none;
    border-radius: 2em;
    background-color: #090;
}
```

You can see that by default, `callButton` is not visible. We do this when the user does not see the local video yet. We will show the button once the local video is established.

It is all about HTML layout and styles. Now, let's work on the actual WebRTC logic.

In the `index.js` file we start with a script similar to what we used in the previous application `CrazyBubbles`:

```
var isBrowser = document.URL.indexOf( 'http://' ) !== -1 || document.
URL.indexOf( 'https://' ) !== -1;
if ( !isBrowser ) {
    document.addEventListener('deviceready', init, false);
} else {
    init();
}
```

With this code, we checked whether the application is running in the browser, or whether it is an actual mobile application and is running on a mobile device. If there is no HTTP or HTTPS in the document URL, then we would interpret it as a mobile application. Otherwise, it is running in a web browser. If it is a mobile application, we would attach the `init()` function as an event handler for the `deviceready` event. If the application is running in a web browser, then we would simply call it the `init()` function.

The init() function is just a wrapper for all our internal functions. In the beginning of the function, there are basic definitions of the required elements:

```
function init() {

    var PeerConnection = window.mozRTCPeerConnection ||
    window.webkitRTCPeerConnection;
    var IceCandidate = window.mozRTCIceCandidate ||
    window.RTCIceCandidate;
    var SessionDescription = window.mozRTCSessionDescription ||
    window.RTCSessionDescription;
    navigator.getUserMedia = navigator.getUserMedia ||
    navigator.mozGetUserMedia || navigator.webkitGetUserMedia;

    var pc; // PeerConnection

    var SERVER_IP = '192.168.0.102';
    var SERVER_PORT = '1234';

    // DOM elements manipulated as user interacts with the app
    var callButton = document.querySelector("#callButton");
    var localVideo = document.querySelector("#localVideo");
    var remoteVideo = document.querySelector("#remoteVideo");

    callButton.addEventListener('click', createOffer);

    // ...
}
```

The first four rows are mainly used to get proper instance of the required component in different browsers. They are currently prefixed in all browsers. So, we should include polyfill.

Where:

- PeerConnection is a WebRTC component that handles connection and data streaming between two peers

- IceCandidate is a remote candidate to the ICE agent

- SessionDescription is returned by PeerConnection; it can be local and remote

- navigator.getUserMedia represents synchronized streams of media, from the camera and microphone in our case

- SERVER_IP is an IP address of the server where STUN is hosted; it should be the same as we defined on the server side

- `SERVER_PORT` is a port where the STUN server is hosted; it should be the same as we defined on server side
- `callButton` is a button on the page to initiate the call
- `localVideo` is an HTML5 video tag to display your own video
- `remoteVideo` is an HTML5 video tag to display the video from the remote peer

The last string contains a click event handler for the call button. Once the button is clicked, we will call the `createOffer` function, which we will describe later.

After that, we will get the local audio and video:

```
navigator.getUserMedia({
        audio: true,
        video: true
    },
    gotStream,
    function(error) {
        console.log(error)
    }
);
```

The `getUserMedia` function is a part of the `MediaStream` API. `MediaStream` has an input as well as an output, which we can pass to a video element or `RTCPeerConnection`.

The `getUserMedia` function takes the following three parameters:

- The first parameter is constraints; we can specify what we want to receive from the input of `MediaStream`
- The second parameter is a success callback where we receive a stream and pass it to the `gotStream` function
- The third parameter is a failure callback

Once we receive the stream from the local camera and microphone, we will display a call button, assign the video stream to the local video element, and create a new peer connection with the stream:

```
function gotStream(stream) {

    callButton.style.display = 'block';
    localVideo.src = URL.createObjectURL(stream);
```

```
        pc = new PeerConnection(null);
        pc.addStream(stream);
        pc.onicecandidate = gotIceCandidate;
        pc.onaddstream = gotRemoteStream;
    }
```

Where:

- URL.createObjectURL creates DOMString containing a URL, which represents the stream object in parameter
- new PeerConnection(null) creates a peer connection
- pc.addStream adds MediaStream from the local camera and microphone to the peer connection
- the pc.onicecandidate handler is run when a network candidate becomes available
- the pc.onaddstream handler is run when the remote stream is received

In the gotIceCandidate handler, we receive events, checking whether there is a candidate and sending a message back with several options:

```
function gotIceCandidate(event) {
    if (event.candidate) {
        sendMessage({
            type: 'candidate',
            label: event.candidate.sdpMLineIndex,
            id: event.candidate.sdpMid,
            candidate: event.candidate.candidate
        });
    }
}
```

In sendMessage, we pass an object with the following properties:

- type defines the type of the message
- label is the index of m-line in the SDP that this candidate is associated with
- id contains the identifier of "media stream identification"
- candidate carries the candidate-attribute

When we receive remote stream in gotRemoteStream handler we simply create an object URL from the stream and assign it to the remote video element.

```
function gotRemoteStream(event) {
    remoteVideo.src = URL.createObjectURL(event.stream);
}
```

Now let's look into `createOffer()` function which will be called once the user clicks on the call button. Inside we call peer connection method `createOffer`:

```
function createOffer() {
    pc.createOffer(
        gotLocalDescription,
        function(error) {
            console.log(error)
        }, {
            'mandatory': {
                'OfferToReceiveAudio': true,
                'OfferToReceiveVideo': true
            }
        }
    );
}
```

Here, `createOffer` generated a blob that contains an offer with the supported configurations for the session and takes the following three parameters:

- **Success callback**: This receives local description
- **Failure callback**: This shows the error in the console
- **Constraints**: We set mandatory offers to receive audio and video

Similarly, this can be done with `createAnswer`:

```
function createAnswer() {
    pc.createAnswer(
        gotLocalDescription,
        function(error) {
            console.log(error)
        }, {
            'mandatory': {
                'OfferToReceiveAudio': true,
                'OfferToReceiveVideo': true
            }
        }
    );
}
```

You can see that we used `gotLocalDescription` for the success callback. This function receives a description. We set this description as local for the peer connection and send a message to the server with the description:

```
function gotLocalDescription(description) {
    pc.setLocalDescription(description);
    sendMessage(description);
}
```

Finally, we have to implement the `socket.io` functions. Let's create a connection using predefined server IP addresses and port values:

```
var socket = io.connect(SERVER_IP, {
    port: SERVER_PORT
});
```

Now, we have a socket instance and can use it to send and receive messages. You already saw the usage of the `sendMessage` function earlier in this chapter, but an implementation of the function is just one line of code:

```
function sendMessage(message) {
    socket.emit('message', message);
}
```

In `sendMessage`, we sent messages to the server, where they are sent as broadcast messages.

Now, we need to listen for three different message types from the server: `offer`, `answer`, and `candidate`:

```
socket.on('message', function(message) {
    if (message.type === 'offer') {
        pc.setRemoteDescription(new SessionDescription(message));
        createAnswer();
    } else if (message.type === 'answer') {
        pc.setRemoteDescription(new SessionDescription(message));
    } else if (message.type === 'candidate') {
        var candidate = new IceCandidate({
            sdpMLineIndex: message.label,
            candidate: message.candidate
        });
        pc.addIceCandidate(candidate);
    }
});
```

If we receive a message with the `offer` type, we understand that it is an inbound call. We then create `SessionDescription` with message content, set it as remote description for the peer connection, and create an answer.

If it is an answer we do the same, but do not create answer.

For the candidate message type, we created `IceCandidate` and added it to the peer connection. Each media stream has a label, such as `Gk7EuLhsuPTbnjFGkR7xPPJK8ONsgwFFvRS`.

It is all about code in `index.js`. Now, we need to do some tweaks to the mobile application code.

Cordova application tweaks

To give access to the camera and microphone on the mobile device, we should add the following code to `platforms/android/AndroidManifest.xml`:

```
<uses-permission android:name="android.permission.RECORD_AUDIO" />
<uses-permission android:name="android.permission.CAMERA" />
```

However, you may realize that audio from the mobile device will not pass anyway. There is another plugin to help with it. It is `org.chromium.audiocapture`. This plugin sets permissions to capture audio from the device's microphone via the `getUserMedia` API. We can install this plugin with the following command:

```
$ cordova plugin add org.chromium.audiocapture
```

Running the application

The application development is finished, and we need to start all parts of it and try to call.

Start the signaling server:

```
$ cd server
$ node server.js
```

Start the client in the browser:

```
$ cd client/www
$ python -m SimpleHTTPServer 8000
```

Start the mobile application on the real device:

```
$ cd client/www
$ cordova run android
```

Now, when we open `http://localhost:8000` in the browser, we will be requested to give access to the camera and microphone:

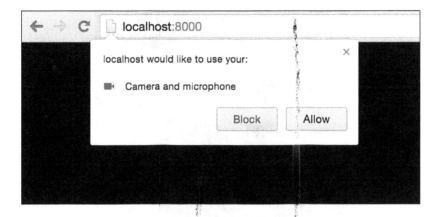

Let's click on **Allow**, and we will see something like the following screenshot:

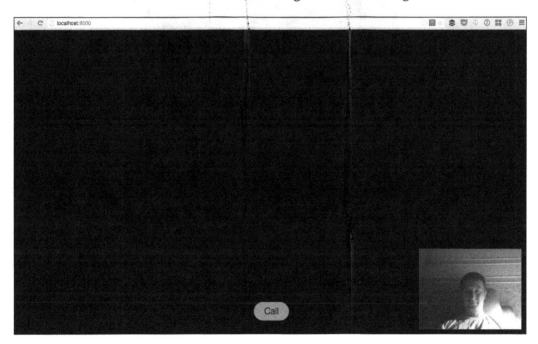

We will see a similar screen on the mobile device. However, it will not request access to the camera and microphone.

Here is the most interesting part. When we click on the **Call** button in a mobile application or in a browser, it doesn't matter. We will be able to see the established call between two endpoints.

The following is the established call on the desktop:

And the following is the established call on the mobile:

The video is working, and the audio is passing in both directions. Congratulations! We have successfully created our first WebRTC application with Cordova/ PhoneGap and Node.js.

Building a real-time communication application with PeerJS

Now, we can do the same, but with the help of the PeerJS library. In the previous example, you may notice that we can do a call only between two peers. In the following example, we will show you how we can initiate a call between more than two peers.

PeerJS is a wrapper for the browser's WebRTC implementation. It is aimed to simplify the peer-to-peer connection management. PeerJS provides a functionality to list the connected clients.

Server side

We will create a simple Node.js application with peer module. For example, we will add two modules: peer and ip. We can do this with the following commands:

```
$ npm install peer --save
$ npm install ip --save
```

After this installation, the package.json file might look like this:

```
{
  "name": "pumpidu-peerjs",
  "version": "0.0.0",
  "description": "Run a PeerJS WebRTC server",
  "main": "index.js",
  "scripts": {
    "test": "echo \"Error: no test specified\" && exit 1"
  },
  "author": "Andrew Kovalenko <cybind@gmail.com>",
  "license": "MIT",
  "dependencies": {
    "peer": "^0.2.5",
    "ip": "^0.3.0"
  }
}
```

Once we create the `server.js` file, we will be able to see the following folder structure:

```
.
├── node_modules
│   ├── ip
│   └── peer
├── package.json
└── server.js
```

Let's add some logic to `server.js`.

First of all, we will require `peer` and `ip`, and create a peer server:

```
var ip = require('ip');
var port = 9000;
var PeerServer = require('peer').PeerServer;
var server = new PeerServer({
    port: port,
    allow_discovery: true
});
```

Here, `port` defines the port number where the peer server will be running, and the `allow_discovery` property is needed to let people who are connected see each other.

Now, we will attach only two handlers, to connect and disconnect:

```
server.on('connection', function(id) {
    console.log('new connection with id ' + id);
});

server.on('disconnect', function(id) {
    console.log('disconnect with id ' + id);
});
```

We do this only to display information for us. Here is a little more information for us once the server has started:

```
console.log('peer server running on ' + ip.address() + ':' + port);
```

Really, we just developed a custom server with the PeerJS server library. So, to start it, we can run using the following command:

```
$ node server.js
```

The server will be running on port 9000 in this case. However, we can do this in an easier way and just run the PeerJS server in the command line:

```
$ peerjs --port 9000 --key peerjs
```

So, we got the server installed and configured it. Now, we can work on the client part.

Client side

We keep all the CSS and HTML the same as in previous chapter. We only want to change the socket.io library to peer.js, so that it looks like this:

```
<script type="text/javascript" src="cordova.js"></script>
<script type="text/javascript" src="js/peer.js"></script>
<script type="text/javascript" src="js/index.js"></script>
```

 We took the PeerJS client library from http://peerjs.com/.

The header of the index.js file is the same as in the previous example. We only change the content of the init() function. At the top of the function, we will define the following variables:

```
var SERVER_IP = '192.168.0.102';
var SERVER_PORT = 9000;

var callButton = document.querySelector("#callButton");
var localVideo = document.querySelector("#localVideo");
var remoteVideo = document.querySelector("#remoteVideo");

var callerId = null;
var peer = null;
var localStream = null;
```

Where:

- SERVER_IP and SERVER_PORT is a PeerJS server location; I used IP addresses from the local network, so I can make calls between the mobile device and browser on the desktop in the same network

- callButton, localVideo, and remoteVideo are DOM elements manipulated as the user interacts with the app

- callerId is the ID set for the current client

- `peer` is the PeerJS object instantiated when the current client connects with its caller ID
- `localStream` is the local video stream captured with `getUserMedia()`

Once the application starts after the initialization of the preceding variables, we will display the prompt window to enter our caller ID (nickname):

```
setCallerId();
```

In this function, we will simply show the prompt with the request and try to initiate a connection after that:

```
var setCallerId = function () {
    callerId = prompt('Please enter your name');
    connect();
};
```

There are several code blocks in the `connect()` function.

First, we will check whether the caller ID is set:

```
if (!callerId) {
    alert('Please enter your name first');
    setCallerId();
    return;
}
```

If it is not set, then we will try to assign it again.

We wrap the actual connection to the PeerJS server with the try-catch block:

```
try {

    console.log('create connection to the ID server');
    console.log('host: ' + SERVER_IP + ', port: ' + SERVER_PORT);
    peer = new Peer(callerId, {
        host: SERVER_IP,
        port: SERVER_PORT
    });

    // ...
} catch (e) {
    peer = null;
    alert('Error while connecting to server');
}
```

Here, in the `peer` constructor, we passed the caller ID and server parameters: IP and port. If the connection is created successfully, we would assign the `onclose`, `onopen`, and `on call` handlers:

```
peer.socket._socket.onclose = function() {
    alert('No connection to server');
    peer = null;
};
```

It is hacked to get around the fact that if a server connection cannot be established, the peer and its `socket` property both still have `open` equal true. Instead, listen to the wrapped `WebSocket` and show an error if its `readyState` becomes `CLOSED`:

```
peer.socket._socket.onopen = function() {
    getLocalStream(function() {
        callButton.style.display = 'block';
    });
};
```

Here, we got the local stream ready for incoming calls once the wrapped `WebSocket` was open. Eventually, we will handle events representing the incoming calls:

```
peer.on('call', answer);
```

As you can see, when the connection is established, we will call the `getLocalStream` function:

```
var getLocalStream = function(successCb) {
    if (localStream && successCb) {
        successCb(localStream);
    } else {
        // ...
    }
};
```

Here, `successCb` has the `successCb(stream)` signature, and receives the local video stream as an argument.

The `main` function we use inside is the same as in the previous example:

```
navigator.webkitGetUserMedia({
        audio: true,
        video: true
    },
    function(stream) {
        localStream = stream;
```

```
        localVideo.src = window.URL.createObjectURL(stream);
        if (successCb) {
            successCb(stream);
        }
    },
    function(err) {
        alert('Failed to access local camera');
        console.log(err);
    }
);
```

Here, we got the local video and audio stream and showed the preview in the local video element. Also, we display the call button once we get media streams.

Once we get an incoming call, we will execute the following function:

```
var answer = function(call) {
    if (!peer) {
        alert('Cannot answer a call without a connection');
        return;
    }
    if (!localStream) {
        alert('Could not answer call as there is no localStream
        ready');
        return;
    }
    console.log('Incoming call answered');
    call.on('stream', showRemoteStream);
    call.answer(localStream);
};
```

We did several verifications here. Peer and `localStream` are both created. After that, we will attach a handler on the `stream` event to the `call` object where we just display a remote stream:

```
var showRemoteStream = function(stream) {
    remoteVideo.src = window.URL.createObjectURL(stream);
};
```

Also, we executed the answer method of the incoming `call` object to bind it to the local stream.

That is all about inbound calls. Now, we need to write some logic for outbound calls.

Let's attach the `dial()` function as an event handler for the call button click:

```
callButton.addEventListener('click', dial);
```

In the `dial()` function, we will do several checks as well:

```
if (!peer) {
    alert('Please connect first');
    return;
}

if (!localStream) {
    alert('Could not start call as there is no local camera');
    return
}
```

Here, we checked whether `peer` and `localStream` are both present. The show prompt window is used to capture another peer's ID (nickname):

```
var recipientId = prompt('Please enter recipient name');

if (!recipientId) {
    alert('Could not start call as no recipient ID is set');
    dial();
    return;
}
```

Here, we checked whether `recipientId` is empty, and if it is empty, we will enter it once more.

After that, we will get a local stream and initiate the call using `recipientId` that we get from the prompt:

```
getLocalStream(function(stream) {
    console.log('Outgoing call initiated');
    var call = peer.call(recipientId, stream);
    // ...
});
```

We would add two handlers to capture the remote stream in response. It would throw an error if something went wrong:

```
call.on('stream', showRemoteStream);
call.on('error', function(e) {
    alert('Error with call');
    console.log(e.message);
});
```

Now, we can run our application.

Running the application

To launch all parts of our application, we will do everything in the same way as in the previous application without PeerJS.

Start the signaling server:

```
$ cd server
$ node server.js
```

Start the client in the browser:

```
$ cd client/www
$ python -m SimpleHTTPServer 8000
```

Start a mobile application on the real device:

```
$ cd client/www
$ cordova run android
```

Now, let's open `http://localhost:8000` in the browser.

We will be requested to enter our client ID:

After we enter our nickname and click on **OK**, we are requested to give access to the camera and microphone:

We are able to see the local video now:

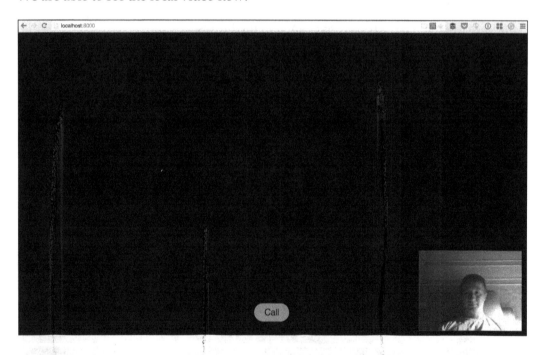

Now, we can look at the mobile application and do everything in the same way.

Enter your own nickname:

Wait while the local media streams initiate:

After that, we will initiate the call by clicking on the **Call** button. We will see a prompt to enter the recipient's nickname. Let's add a nickname for the peer in the desktop's browser:

Then, click on the **OK** button in the prompt. Right after that, we will see the call initiated:

Congratulations! We have built a peer-to-peer WebRTC application with the PeerJS library. Now, we can connect to different peers by only knowing their nicknames.

Exploring other tools to build WebRTC mobile applications

As you can see, we developed a WebRTC application only for desktop browser and Android. Unfortunately, web technologies are better implemented in the browsers than in `WebView`. I hope at the nearest time the iOS `WebView` will support WebRTC, or we will be able to use Crosswalk on the iOS.

However, there are several solutions available. Let's look at a few of them.

OpenTok

OpenTok is complex paid solution that helps add video/audio calls to the mobile applications or to the website. You can see a detailed description, and the price on their website at `https://tokbox.com/`. In this case, you do not need to set up a server. It is just a service that provides the STUN and TURN functionality for you. Tariffs are based on call durations.

 There is Cordova plugin for the service as well. You can find it at `https://github.com/songz/cordova-plugin-opentok`.

Android installation is pretty easy, but iOS configuration needs some effort.

PhoneRTC

It is a PhoneGap/Cordova plugin, and it is free.

 You can find more about it at `http://phonertc.io/`.

I had no problem installing it for Android, but again, I faced some difficulties to set it up for iOS.

Also, you can learn how the PhoneRTC plugin worked with `SIP.js`.

`SIP.js` is a simple, intuitive, and powerful JavaScript signaling library

The plugin can be found at `https://github.com/onsip/onsip-cordova`.

In both OpenTok and PhoneRTC, Cordova plugins use native implementation of WebRTC for Android and iOS. It does not go through `WebView`.

You can see native WebRTC codes and usage at:

- `http://www.webrtc.org/native-code/android`
- `http://www.webrtc.org/native-code/ios`

Summary

In this chapter, you learned what is WebRTC, how it works, and the greatness of this new technology. We explored the main components of WebRTC and the main advantages in using it with PhoneGap/Crodova. Also, we developed a custom server to handle peer-to-peer connection in two different ways. Node.js with its modules handles both the ways pretty easily. As a result, we successfully created two versions of the WebRTC application.

In the next chapter, we will write an Instagram-like application to apply effects to our pictures. We will revisit the application structure with Sencha Touch, process images with HTML5 Canvas and the Pixastic library, and build a custom PhoneGap/Cordova plugin.

8

Building "Imaginary" – An Application with Instagram-like Image Filters

In the previous chapter, we implemented the WebRTC PhoneGap/Cordova application. We built a custom STUN (signaling) server with Socket.io, which processes a variety of requests. Also, we used the PeerJS tool to simplify peer-to-peer connections. Most of the chapters in this book are related to Web development, and only a little bit describes platform-specific features. It's interesting that with the help of PhoneGap/Cordova, we can implement native components and connect them as PhoneGap/Cordova plugins to the application. In this chapter, we will write an Instagram-like application to apply effects to our pictures.

In this chapter, we will:

- Revisit application structure organization with Sencha Touch
- Understand image processing with HTML5 Canvas and the Pixastic library
- Resize images with the help of HTML5 Canvas
- Build a custom PhoneGap/Cordova plugin for iOS

I assigned the name "Imaginary" to the project.

An overview of the Pixastic library

We would like to easily apply image filters to the pictures we get from the camera on our device. After some research on the Web, I was able to find an awesome library named Pixastic.

 You can download the latest library files from the GitHub repository at https://github.com/jseidelin/pixastic.

It uses HTML5 Canvas for image processing. There are different scenarios for the library usage, but I selected the one with web workers. A web worker is a JavaScript code running in the background without blocking the web page and without affecting the performance of the page. For me, it seems to work best for mobile devices. There are four main files needed:

- `pixastic.js`: This is the main file with basic logic
- `pixastic.effects.js`: This file affects logic and pixel processing
- `pixastic.worker.control.js`: This is used as the control of the worker
- `pixastic.worker.js`: This is the worker file itself

Here is a simple example of the Pixastic library usage, which we will follow in this chapter:

```
var img = new Image();
img.onload = function () {

    var oc = document.createElement('canvas'),
        octx = oc.getContext('2d');

    oc.width = img.width;
    oc.height = img.height;

    octx.drawImage(img, 0, 0, img.width, img.height);

    P = new Pixastic(octx);
    P['mosaic']({ blockSize: 8 }).done(function() {
        // processing finished
        var data = oc.toDataURL();
    }, function(p) {
        // display progress here;
    });

};

img.src = 'some/image/url/image.jpg';
```

Where we create the `Image` object, we assign a source to it and wait for the picture to load. After this, we create a `canvas` element on the page and take its context with the `oc.getContext('2d')` function.

 2D context provides objects, methods, and properties to draw the painting area and perform other object manipulations.

Then, we assign the width and height of our picture to the canvas. We draw an image on the canvas with the `drawImage(img, 0, 0, img.width, img.height)` function. This function has the following properties:

- The first argument is the `Image` object
- The second and third arguments are coordinates where we will place the image on the canvas
- The fourth and fifth arguments are the size of the image to use (stretch or reduce the image)

After that, we will create a Pixastic object where we pass canvas's `new Pixastic(octx)` context. Now, we can process the `P['mosaic']({ blockSize: 8 }).done(a,b)` effect.

Where:

- `mosaic` is the effect name
- `{ blockSize: 8 }` is the options for the effect
- `a` is a success callback
- `b` is a fail callback

Once processing is finished, we can get the picture with effect by simply calling `oc.toDataURL()`. It will return Base64-coded image data, which we can use.

So, we reviewed the Pixastic library a little. Let's jump into the development of the Imaginary application.

Bootstrapping the Sencha Touch application

We will start by creating a project with the Sencha Touch Cmd. Let's generate the application (we already learned this in *Chapter 2, Setting Up a Project Structure with Sencha Touch*):

```
$ cd /Development/touch-2.4.1
$ sencha generate app Imaginary ~/Projects/phonegap-by-example/imaginary
```

I will not explain what the parameters mean here because we already looked into it. The initial folders and files structure looks like this:

```
.
├── app
├── app.js
├── app.json
├── bootstrap.js
├── bootstrap.json
├── build
├── build.xml
├── cordova
├── index.html
├── packages
├── resources
└── touch
```

Also, we will install several Cordova plugins:

```
$ cd cordova
$ cordova plugin add org.apache.cordova.statusbar,
$ cordova plugin add org.apache.cordova.camera
$ cordova plugin add org.apache.cordova.file
```

Each of them is responsible for status bar display, camera access, and filesystem access respectively. We will use these plugins in the upcoming chapters.

We only add configuration for the status bar. We add it before closing the widget section in the `cordova/config.xml` file:

```
<preference name="StatusBarOverlaysWebView" value="false" />
<preference name="StatusBarBackgroundColor" value="#000000" />
<preference name="StatusBarStyle" value="lightcontent" />
```

The preceding code just defines the color of the status bar, its overlay web view, and its style.

In the `app.json` config section, we specify our targeting platform, `"platforms"`: `"ios"`. In `app.js`, we add a single view and controller:

```
views: [ 'Main' ],
controllers: [ 'Main' ]
```

The `Main.js` view represents the Sencha Touch panel component with `TitleBar` and tabs at the bottom:

```
Ext.define('Imaginary.view.Main', {
    extend: 'Ext.tab.Panel',
    xtype: 'main',
    requires: [
        'Ext.TitleBar',
        'Ext.Button',
        'Ext.Img'
    ],
    config: {
        tabBarPosition: 'bottom',
        items: [
            //...
        ]
    }
});
```

Let's define two different tabs. The first one will contain the **Take Photo** button. When we click on the button, we should be able to see the camera popup screen. We will place this button in the middle of the container:

```
{
    title: 'New Photo',
    iconCls: 'lens',
    items: [
        {
            docked: 'top',
            xtype: 'titlebar',
            title: 'New Photo'
        },
        {
            xtype: 'container',
            width: '100%',
            height: '100%',
            layout: {
```

```
                    type: 'vbox',
                    pack: 'center',
                    align: 'center'
                },
                items: [
                    {
                        xtype: 'button',
                        id: 'takePhotoBtn',
                        text: 'Take Photo',
                        iconCls: 'photo',
                        iconAlign: 'top',
                        height: 70,
                        width: 120,
                        padding: 10,
                        margin: 5
                    }
                ]
            }
        ]
    }
```

As you can see, there are `iconCls` with the value `photo` assigned to the button. It
is the `icon` class we define in `resources/sass/app.scss`. We add some other icon
class's definitions as well to the button:

```
@include icon('lens', 'L');
@include icon('photo', 'v');
@include icon('globe', 'G');
@include icon('check', '3');
@include icon('gallery', 'P');
```

On the second tab, we define only an empty container to display the pictures we will
capture with the camera:

```
{
    title: 'My Photos',
    iconCls: 'gallery',
    items: [
        {
            docked: 'top',
            xtype: 'titlebar',
            title: 'My Photos'
        },
```

```
        {
            xtype: 'container',
            id: 'photos',
            width: '100%',
            height: '100%',
            scrollable: {
                direction: 'vertical',
                directionLock: true
            }
        }
    ]
}
```

You can see that we created the container with `100` percent width and height and made it vertically scrollable.

There is not much inside the `Main.js` controller. We added only two references to the buttons and container for pictures:

```
Ext.define('Imaginary.controller.Main', {
    extend: 'Ext.app.Controller',
    config: {
        refs: {
            takePhotoBtn: '#takePhotoBtn',
            photoContainer: '#photos'
        },
        control: {
            //...
        }
    },
    //...
});
```

Now, we can see our initial application by running this command:

```
$ sencha app build -run native
```

It will show us the following screen:

And that is it! Now, let's go to the photo capture implementation.

Capturing photos

To capture a photo in the `Main.js` controller, we need to add a click handler to the **Take Photo** button. We can do this by adding such a tap handler in the control section:

```
control: {
    takePhotoBtn: {
        tap: 'getPhoto'
    }
}
```

After that, we define a variable in the controller to store the link to the original image without any applied effect:

```
originalImageUri: null
```

When the user taps the **Take Photo** button, the getPhoto function will be called. In the function, we take pictures and assign its URL to the originalImageUri controller's variable to reference later:

```
getPhoto: function() {
    var self = this;
        self.getCameraPicture(function(imageURI) {
         self.originalImageUri = imageURI;
         self.showPhotoPopup(imageURI);
    });

}
```

You can see that here, we called the self.getCameraPicture function, where we access the camera plugin:

```
getCameraPicture: function(callback) {
    if (Ext.browser.is.PhoneGap) {
            var onSuccess = function(imageURI) {
             if (callback) callback(imageURI);
        }
        var onFail = function(message) {
            alert('Failed because: ' + message);
            if (callback) callback();
        }
          navigator.camera.getPicture(onSuccess, onFail, {
          quality: 50,
          targetWidth: 1000,
          targetHeight: 1000,
          destinationType:
          navigator.camera.DestinationType.FILE_URI,
          mediaType: navigator.camera.MediaType.PICTURE,
          sourceType: navigator.camera.PictureSourceType.CAMERA,
          encodingType: navigator.camera.EncodingType.JPEG,
          correctOrientation: true,
          saveToPhotoAlbum: false
        });
    } else {
        // Emulate captured image
        if (callback) callback('resources/images/test.jpg');
    }
}
```

We are checking here whether the application is running in the browser or as a native on the device. If it is browser, then we emulate the captured image and simply return the URL to the local test image. If it is running as a native on the device, we would call the `navigator.camera.getPicture` function. We described all the properties of this function in *Chapter 3, Easy Work with Device – Your First PhoneGap Application "Travelly"*.

In this case, we reduce the image size to be not more than 1,000 pixels width and height. We return the file URI, not the Base64 string so that we can reference it with the link to the filesystem. If the picture was captured successfully, we would pass the file URI in the callback function. In this case, we just call the callback without arguments.

Rendering an effects list

We would like to display pictures captured with a camera in the new popup and display the list of available effects as a preview of the captured picture. However, before this, we need to include the Pixastic library first.

Including Pixastic

Let's download the Pixastic files and put them in the `resources/lib` folder. After that, we need to include these files in our project. We can do this in the `app.json` file in the `js` section:

```
"js": [
    {
        "path": "resources/lib/pixastic.js"
    },
    {
        "path": "resources/lib/pixastic.effects.js"
    },
    {
        "path": "resources/lib/pixastic.worker.js"
    },
    {
        "path": "resources/lib/pixastic.worker.control.js"
    }
    //...
]
```

The preceding code tells the Sencha microloader to include these files in `index.html`. Also, we need to specify the `resources/lib` folder as a resource itself. So, it will be copied in the application as well. We need it to reference from the code later, as shown in the following code:

```
"resources": [
    //...
    "resources/lib"
]
```

Now, we will display the popup with captured picture.

showPhotoPopup

Earlier in the `getPhoto` function, we called the `self.showPhotoPopup(imageURI)` function. This function has to create another panel and show it on top of the main view. Let's take a closer look at the function.

```
showPhotoPopup: function(imageURI) {
    var self = this;
    var popup = Ext.create('Imaginary.view.NewPicture');
    Ext.Viewport.add(popup);
    popup.show();

    //...

    self.setPreviewImage(imageURI);
    popup.on('hide', function() {
        popup.destroy();
    });
    Ext.getCmp('retakePhotoBtn').on('tap', function() {
        self.getCameraPicture(self.setPreviewImage);
    });
    Ext.getCmp('savePhotoBtn').on('tap', function() {
        //...
    });
    Ext.getCmp('cancelPhotoBtn').on('tap', function() {
        popup.hide();
    });
}
```

In the function, we created an instance of the `Imaginary.view.NewPicture` view, added it to the viewport, and displayed it. The view itself looks like this:

```
Ext.define('Imaginary.view.NewPicture', {
    extend: 'Ext.Panel',
    xtype: 'newpicture',
    requires: [
        'Ext.TitleBar',
        'Ext.Button',
        'Ext.Img'
    ],
    config: {
        height: '100%',
        width: '100%',
        centered: true,
        showAnimation: 'slideIn',
        hideAnimation: 'slideOut',
        hidden: true,
        items: [
            //...
        ]
    }
});
```

It is a full-screen panel that appears with the slide-in effect and closes with the slide-out effect. By default, we keep it hidden. The items are `titlebar` and full screen container, as shown in the following code:

```
{
    docked: 'top',
    xtype: 'titlebar',
    title: 'New Photo'
},
{
    xtype: 'container',
    width: '100%',
    height: '100%',
    layout: {
        type: 'vbox',
        pack: 'center',
        align: 'center'
    },
    items: [
        //...
    ]
}
```

As you can see, here we added the layout configuration so that all elements are aligned vertically in the center.

In the container, let's add our main components: image preview, a container for the effects list, and action buttons:

```
{
    xtype: 'image',
    id: 'photoPreview',
    width: '100%',
    flex: 8
},
{
    xtype: 'container',
    id: 'effectsContainer',
    flex: 1,
    width: '100%',
    layout: {
        type: 'hbox',
        pack: 'center',
        align: 'center'
    },
    scrollable: {
        direction: 'horizontal',
        directionLock: true
    }
},
{
    xtype: 'container',
    flex: 1,
    layout: {
        type: 'hbox',
        pack: 'center',
        align: 'center'
    },
    items: [
        {
            xtype: 'button',
            id: 'retakePhotoBtn',
            text: 'Retake',
            flext: 1,
            margin: '0 5 0 5'
        },
        {
```

```
                    xtype: 'button',
                    id: 'savePhotoBtn',
                    text: 'Save',
                    flext: 1,
                    margin: '0 5 0 5'
            },
            {
                    xtype: 'button',
                    id: 'cancelPhotoBtn',
                    text: 'Close',
                    flext: 1,
                    margin: '0 5 0 5'
            }


        ]
    }
```

As you can see, for each component, we assigned an ID so that we can reference them from our controller. Now, let's go back to the popup implementation in the controller.

Once we have displayed the popup, we will assign the `self.setPreviewImage(imageURI)` function to the picture. This function references our Sencha Touch Image component with the `photoPreview` ID in the popup and assigns its source:

```
setPreviewImage: function(imageURI) {
    Ext.getCmp('photoPreview').setSrc(imageURI);
}
```

After that, we define several event handlers to:

- Hide the popup
- Retake, save, and cancel the button's clicks

Once the popup is hidden, we just destroy the object in the memory to improve the application's performance and memory utilization.

Also, we are handling tap events on these three buttons. Once the user taps on the retake button, we will call the `self.getCameraPicture(self.setPreviewImage)` function, which is already defined where the callback is assigned to the newly captured picture to preview. When the user taps on the **Cancel** button, we simply hide the popup. If the user taps the **Save** button, we would store the picture.

However, before defining the save function, let's implement the effects list first.

Defining the effects model and store

To properly work with the list of effects, we will define the effect model and store first.

As we already know that the effect has `name` and `options`, the model looks pretty simple:

```
Ext.define('Imaginary.model.Effect', {
    extend: 'Ext.data.Model',
    config: {
        fields: [
            { name: 'name', type: 'string' },
            { name: 'options', type: 'auto' }
        ]
    }
});
```

It has only two fields: `name` and `options`. The `name` property is assigned a `string` type, and we assigned the `auto` type to `options` to allow us to store custom JSON data in it. I will add this model in the `app/model/Effect.js` file.

The store is simple as well. I just took at a list of all the available effects for the Pixastic library and created a store in the `app/store/Effects.js` file with inline data. Here is an example only with few of the effects:

```
Ext.define('Imaginary.store.Effects', {
    extend: 'Ext.data.Store',
    requires: ['Imaginary.model.Effect'],
    config: {
        storeId: 'Effects',
        model: 'Imaginary.model.Effect',
        data: [
            { name: 'posterize', options: { levels: 5 } },
            { name: 'solarize' },
                //..
        ]
    }
});
```

In total, there are 29 effects.

Applying effects to thumbnails

Now, we can loop through effects in the store and create thumbnails for them. Let's define the function to resize the image:

```
resizeImage: function(imageURI, callback) {
    var img = new Image();
    img.onload = function () {
        var oc = document.createElement('canvas'),
            octx = oc.getContext('2d');
        var ratio = 160/img.width;
        var newWidth = img.width*ratio;
        var newHeight = img.height*ratio;
        oc.width = newWidth;
        oc.height = newHeight;
        octx.drawImage(img, 0, 0, newWidth, newHeight);
        if (callback) callback(oc.toDataURL(), oc, octx);
    };
    img.src = imageURI;
}
```

In the function, we created a new `Image` object based on the URI we get. After that, we loaded it. Here, we used the HTML5 Canvas element to resize the image. We created the canvas and took its 2D context for drawing. I would like the image to not be wider than 160 pixels. I calculated the ratio based on this and detected a new image size, which I assigned to the canvas. Then, I drew the image with a new size on the canvas and simply returned a Base64 representation of the image. In this case, the image will be scaled down to 160 pixels in width.

We have implemented the function to create thumbnails. Now, we need to implement the function to apply the Pixastic effect. This is also not difficult:

```
applyEffect: function(effect, thumbData, callback) {
    var img = new Image();
    img.onload = function () {

        var oc = document.createElement('canvas'),
            octx = oc.getContext('2d');

        oc.width = img.width;
        oc.height = img.height;

        octx.drawImage(img, 0, 0, img.width, img.height);
```

```
        P = new Pixastic(octx, 'resources/lib/');
        P[effect.name](effect.options).done(function() {
            var data = oc.toDataURL();
            if (callback) callback(data);
        }, function(p) {
            // display progress here;
        });

    };
    img.src = thumbData;
}
```

The preceding code is almost the same as I described earlier in the introduction to Pixastic. The first difference is that we receive the effect name and options as an argument. The second difference lies in passing the second argument in the Pixastic constructor. We need to pass `resources/lib/` for Pixastic to dynamically load the worker control file.

To retrieve the list of effects, we will define the helper function to retrieve them:

```
getEffects: function() {
    var effectsStore = Ext.getStore('Effects');
    var data = effectsStore.getData();
    return data ? data.all : null;
}
```

This simply returns the data option of the store.

Let's look at the actual creation of the effect thumbnails list:

```
showPhotoPopup: function(imageURI) {
    //...
    var effects = self.getEffects();
    var effectsContainer = Ext.getCmp('effectsContainer');
    self.resizeImage(imageURI, function(thumbData) {
        //...
    });
    //...
}
```

As you can see, we added the code to the showPhotoPopup function. We retrieved the list of effects, took the container where we will add the effect thumbnail, and created the thumbnail. In thumbData, we get the minified version of the captured picture. Now, we can loop through the list of effect objects and insert thumbnails in the container:

```
effects.forEach(function(item) {
    var effect = item.getData();
    var effectImage = Ext.create('Ext.Img', {
        src: thumbData,
        height: 64,
        width: 64,
        margin: '2 2 2 2'
    });
    effectImage.on('tap', function() {
        //...
    });
    effectsContainer.add(effectImage);
});
```

As you can see, we created the Ext.Img component, where we assigned thumbData as a source to it and added the image to the container. As you can see, we set the width and height to 64, and it is two times less than the actual thumbnail size. We did this to eliminate pixilation of the thumbnail on high-resolution screens (retina display, 4K displays, and mobile device screens). Also, we added some margin so that the thumbnails are not positioned close to each other. You can see that there is an empty tap event handler. We will define it in the following section.

So, we successfully added thumbnails to the container, and now, we need to loop through them and apply effects to each of them:

```
var thumbs = effectsContainer.getItems();
var i = 0;
var item = thumbs.items[i];
self.applyEffect(effects[i].data, thumbData, function
cb(thumbDataFiltered) {
    if (i < thumbs.items.length-1) {
        item.setSrc(thumbDataFiltered);
        i++;
        item = thumbs.items[i];
        return self.applyEffect(effects[i].data, thumbData, cb);
    }
});
```

In the preceding lines of code, we got a list of already added thumbnails and applied effects using recursion. This allows us to apply effects one after another; it takes some time on the real device. Once the effect is applied, we replace the source of the thumbnail with a new one using the `item.setSrc(thumbDataFiltered)` function.

Once the effects are applied, we will be able to see something like this:

It looks pretty good, right?

Applying effects to the photo

Now, we can implement logic to apply an effect for the big picture we have already displayed. Applying the effect to such a big picture takes some time, so I would like to implement a pre-loader to display it while it is processing. Here is the code I added in the `app/LoadMask.js` file:

```
Ext.define('Imaginary.LoadMask', {
    extend: 'Ext.LoadMask',
    xtype: 'loadmask',

    config: {
        message: '',
      html: '<div class="loader">
      <div class="item-1"><span></span></div>
      <div class="item-2"><span></span></div>
      <div class="item-3"><span></span></div>
      <div class="item-4"><span></span></div>
      <div class="item-5"><span></span></div>
      <div class="item-6"><span></span></div>
      </div>',
      zIndex: 3000
    }
});
```

Here, I left a message blank and added some HTML content, which I will animate with CSS3. Also, I put a big z-index here to be sure that this pre-loader is always displaying on top. Of course, I assigned the class name `Imaginary.LoadMask` to it.

Now, we come back to the logic where we assign thumbnails and tap event handlers:

```
effectImage.on('tap', function() {
    var thumbs = effectsContainer.getItems();
    var index = thumbs.items.indexOf(this);
    popup.setMasked({ xclass: 'Imaginary.LoadMask' });
    self.applyEffect(effects[index].data, self.originalImageUri,
    function(imageDataFiltered) {
        self.setPreviewImage(imageDataFiltered);
        popup.setMasked(false);
    });
});
```

In this tap event handler, we retrieved a list of already added thumbnails and defined the index of the tapped item. After that, we displayed our new pre-loader by calling the `popup.setMasked({ xclass: 'Imaginary.LoadMask' })` function. You can see that here, we specified the name of our pre-loader's class. In this case, it will show custom and nonstandard component.

With the detected index, we get the effect object from the array and push it along with original image data to the `applyEffect` function. Once the picture is processed, we will replace the source of the big picture and hide the pre-loader.

Now, when you click on any thumbnail in the list, you will be able to see the following screenshots:

Congratulations! We successfully applied the effect to the picture. Now, we need to save it.

Saving the dressed photo into the application's folder

To save the picture in the application's folder ,we need to implement the store and model of the picture and some logic.

Defining the picture model and store

The picture model itself is very simple. Let's create it and put it under app/model/Picture.js:

```
Ext.define('Imaginary.model.Picture', {
    extend: 'Ext.data.Model',
    config: {
        fields: [
            { name: 'id', type: 'int' },
            { name: 'url', type: 'string' }
        ]
    }
});
```

Where:

- id is the generated identifier in locaStorage
- url is the path to the picture on the device

The store is a little different than the one for effects. Let's add the following lines of code in the app/store/Pictures.js file:

The Pictures.js store, is as follows:

```
Ext.define('Imaginary.store.Pictures',{
    extend:'Ext.data.Store',
    requires: ['Ext.data.proxy.LocalStorage',
    'Imaginary.model.Picture'],
    config:{
        model:'Imaginary.model.Picture',
        storeId: 'Pictures',
        autoLoad: true,
        autoSync: true,
        proxy:{
            type:'localstorage'
        }
    }
});
```

As a proxy, we use `localStorage` here. The `autoLoad` property means the load method that is called right after the creation of the object. The `autoSync` property specifies automatic sync of the store with its proxy after every edit to one of its records.

Saving the picture to the filesystem

Earlier in the chapter, we described three buttons in the picture preview popup. One of them is the save button and the tap event handler on the button looks like this:

```
Ext.getCmp('savePhotoBtn').on('tap', function() {
    var filteredImageURI = Ext.getCmp('photoPreview').getSrc();
    self.savePhoto(filteredImageURI, function() {
        popup.hide();
    });
});
```

Here, we take the original source of the photo preview and pass it to the `savePhoto` function. In the callback, we simply hid the popup.

Let's take a closer at the `savePhoto` function:

```
savePhoto: function(imageURI, callback) {
    var self = this;
    self.copyPhotoToPersistentStore(imageURI,
    function(persistentImageURI) {
        var picture = Ext.create('Imaginary.model.Picture', {
            url: persistentImageURI
        });
        var pictureStore = Ext.getStore('Pictures');
        pictureStore.add(picture);
        // refresh list of pictures goes here
        if (callback) callback();
    })
}
```

With the help of the `copyPhotoToPersistentStore` function, we saved the image to the filesystem. After that, we created the picture object with a new URL, saving it to the store and executing the callback. I think this is quite clear, but we need to look at what the `copyPhotoToPersistentStore` function does.

The function receives two arguments:

- The first argument is the actual Base64 image data
- The second argument is the callback to call after the processing is finished

The content of the function starts with the definition of variables and the data URI-to-binary conversion:

```
copyPhotoToPersistentStore: function(fileURI, callback) {
    var self = this;
    var d = new Date();
    var n = d.getTime();
    var newFileName = n + ".jpg";
    var myFolderApp = "Imaginary";

    function convertDataURIToBinary(dataURI) {
        var BASE64_MARKER = ';base64,';
        var base64Index = dataURI.indexOf(BASE64_MARKER) +
        BASE64_MARKER.length;
        var base64 = dataURI.substring(base64Index);
        var raw = window.atob(base64);
        var rawLength = raw.length;
        var array = new Uint8Array(new ArrayBuffer(rawLength));

        for (i = 0; i < rawLength; i++) {
            array[i] = raw.charCodeAt(i);
        }
        return array;
    }

    var bin = convertDataURIToBinary(fileURI);
    var bb = new Blob([bin], { type: "image/jpeg" });
    //ask for permission and save logic goes here
}
```

Where:

- `newFileName` is a desired file name based on the current timestamp
- `myFolderApp` is a directory to save the picture
- `convertDataURIToBinary` is a function that converts a Base64 representation of the picture into binary data
- `bin` is a variable where we store binary data
- `bb` is an `image/jpeg` blob that we will use while saving

After that, we will request the filesystem and create a writer. We will use the HTML5 FileSystem API:

```
window.requestFileSystem(LocalFileSystem.PERSISTENT, 0,
function (fileSystem) {
    fileSystem.root.getFile(newFileName, {create: true, exclusive:
    false},
        function (fileEntry) {
            fileEntry.createWriter(function (writer) {
                // write to file code goes here
            }, onError);
        }, onError);
}, onError);
```

Here, in the `requestFileSystem` function, there are the following arguments:

- The first argument (type) defines whether the filesystem is persistent or not. The possible values are `window.TEMPORARY` or `window.PERSISTENT`.

- The second argument is the size (in bytes) the app will require for storage. `0` means maximum.

- The third argument is the success callback.

- The fourth argument is the optional failure callback.

The `fileSystem.root.getFile` function allows us to lookup or create a file that has the following arguments:

- The first argument (`newFileName`) is actually the new file name based on the timestamp we created

- The `create: true` argument creates the file if it doesn't exist

The `createWriter()` method obtains a `FileWriter` object. Once we create the writer, we can write the following lines of code to the file:

```
writer.seek(0);
writer.write(bb);
```

Here, `seek(0)` starts the write position at the beginning of the file, and `write(bb)` actually writes the blob into the file.

When the writing is finished, we can copy the blob to the folder:

```
writer.onwrite = function(e) {
    fileSystem.root.getDirectory(myFolderApp, {
        create: true,
        exclusive: false
    },
    function(directory) {
        fileEntry.copyTo(directory, newFileName, function(entry) {
            if (callback) callback(entry.toURL());
        }, onError);
    },
    onError);
};
```

Here, `getDirectory` looks up for the folder or creates it if it doesn't exist. The `copyTo` method copies `fileEntry` into the `Imaginary` directory. In the end, we returned the path to the image with the `entry.toURL()` function. We save this URL in `localStorage`.

Building a custom plugin to save the picture in the iOS library

We successfully saved the picture into the application's Imaginary folder, but we would like to save the picture with the applied effect to the device's library. In this example, it will be the iOS library.

There is no direct access to the iOS library from the HTML5 FileSystem API. There is no such official plugin. In this case, what we need to do is develop a custom plugin for PhoneGap/Cordova.

I will reference the Cordova plugin development guide and follow all these steps:

You can read in detail about the Cordova plugin development on these websites:

- **Guide**: https://cordova.apache.org/docs/en/5.1.1/guide_hybrid_plugins_index.md.html#Plugin%20Development%20Guide
- **Specification**: https://cordova.apache.org/docs/en/5.1.1/plugin_ref_spec.md.html#Plugin%20Specification

A Cordova plugin is a package with native and JavaScript code that allows WebView to communicate with the native platform on which it runs.

The plugin setup

Let's create a plugin outside our Imaginary project:

```
$ cd ~/Projects
$ mkdir cordova-pugin-imagetolibrary
```

The plugin repository must have a top-level `plugin.xml` manifest file. So, we will create it:

```
$ cd cordova-pugin-imagetolibrary
$ touch plugin.xml
```

We will add the following content to the file:

```xml
<?xml version="1.0" encoding="UTF-8"?>
<plugin xmlns="http://apache.org/cordova/ns/plugins/1.0"
        id="com.cybind.imaginary.imagetolibrary" version="0.0.1">
    <name>ImageToLibrary</name>
    <description>Cordova ImageToLibrary Plugin</description>
    <license>Apache 2.0</license>
    <keywords>cordova,image,base64,library</keywords>
    <engines>
            <engine name="cordova" version=">=3.0.0" />
    </engines>
    <js-module src="www/ImageToLibraryPlugin.js"
    name="imagetolibrary">
        <clobbers target="ImageToLibraryPlugin" />
    </js-module>
    <platform name="ios">
        <config-file target="config.xml" parent="/*">
            <feature name="ImageToLibraryPlugin">
                <param name="ios-package"
                value="ImageToLibraryPlugin" />
                <param name="onload" value="true" />
            </feature>
        </config-file>
        <header-file src="src/ios/ImageToLibraryPlugin.h" />
        <source-file src="src/ios/ImageToLibraryPlugin.m" />
    </platform>
</plugin>
```

Where:

- id is the reverse-domain format string to identify the plugin package
- js-module is the section specifying the path to the common JavaScript interface
- The platform section specifies a set of native code for the iOS platform in this case
- config-file represents the configuration section, which will be inserted into config.xml
- header-file and source-file tags specify the path to the library's component files; in this case, it is Objective-C files

As you can see here, we specified some JavaScript and Objective-C files. Let's put these files in the filesystem so that our plugin's folder looks like this:

```
.
├── plugin.xml
├── src
│   └── ios
│       ├── ImageToLibraryPlugin.h
│       └── ImageToLibraryPlugin.m
└── www
    └── ImageToLibraryPlugin.js
```

Now, let's look at ImageToLibraryPlugin.js.

The JavaScript interface

The JavaScript interface is perhaps the most important part of the plugin. We can add a different structure for the file, but eventually, we have to call cordova.exec to communicate with the native platform. In our case, I did it this way:

```
var exec = require('cordova/exec');
var ImageToLibraryPlugin = {
    saveToLibrary: function(types, success, fail) {
        exec(success, fail, "ImageToLibraryPlugin", "saveImage",
        types);
    }
};
module.exports = ImageToLibraryPlugin;
```

Here, we called the saveToLibrary function, which receives an array with custom data, a success callback, and a failure callback.

The exec Cordova function has the following arguments:

- The first argument is a success callback function.
- The second argument is an error callback function. It returns an optional error parameter.
- The third argument is the service name to call on the native side. Usually, it is a class.
- The fourth argument is the action name to call on the native side. Usually, it is a class method.
- The fifth argument is an array of custom arguments.

Here, we pass an argument array to the `saveImage` method of the `ImageToLibraryPlugin` class in Objective-C.

Native iOS code

Now, let's look at the native implementation of the plugin. I will not go deep into Objective-C development but will describe what I did. We would like to save Base64 image representation into the iOS library so that we can see it outside the application.

There are only two files we defined in `plugin.xml`: `ImageToLibraryPlugin.h` and `ImageToLibraryPlugin.m`.

In the header plugin, we created `ImageToLibraryPlugin` and inherited it from `CDVPlugin`. It is required to make the plugin work, and define the properties and instance methods. So, the header file looks like this:

```
typedef void(^SaveImageCompletion)(NSError* error, NSString* url);

@interface ImageToLibraryPlugin : CDVPlugin
{
    NSString* callbackID;
}

@property (nonatomic, copy) NSString* callbackID;
@property (strong, atomic) ALAssetsLibrary* library;

// Instance Method
-(void)saveImage:(CDVInvokedUrlCommand*)command;
-(void)removeImage:(NSMutableArray*)arguments
withDict:(NSMutableDictionary*)options;
-(void)saveImageToLibrary: (UIImage *)image;
-(void)saveImage:(UIImage*)image toAlbum:(NSString*)albumName withComp
letionBlock:(SaveImageCompletion)completionBlock;
@end
```

The method we will call is `saveImage`. It receives an argument of the `CDVInvokedUrlCommand` type. All other methods are internal, which we use in the native code.

Now, let's look at the implementation file, particularly in the `saveImage` method:

```
-(void)saveImage:(CDVInvokedUrlCommand*)command
{
    self.callbackID = command.callbackId;
    NSString *stringObtainedFromJavascript = [command.arguments
    objectAtIndex:0];

    //Saving image to divice library
    NSURL *url = [NSURL
    URLWithString:stringObtainedFromJavascript];
    NSData *imageData = [NSData dataWithContentsOfURL:url];
    UIImage *image = [UIImage imageWithData:imageData];
    [self saveImageToLibrary:image];
}
```

The first argument in the arguments parameter is `callbackID`. We used this to send data back to `successCallback` or `failureCallback` through `PluginResult`. After that, we got Base64 data passed to the plugin from the JavaScript interface at `[command.arguments objectAtIndex:0]`. Then, we converted Base64 representation to the `UIImage` object and passed it to the `saveImageToLibrary` method.

I will not explain all the logic related to saving the picture in the library; it is beyond the scope of this book. I will show only how to return the plugin result. You can see the actual implementation of the plugin in the GitHub repository.

We need to create an instance of `CDVPluginResult`:

```
CDVPluginResult* pluginResult = nil;
```

Then, we need to assign our data to the plugin result object. If it is successful, we will see the following result:

```
pluginResult = [CDVPluginResult resultWithStatus:CDVCommandStatus_OK
messageAsString: stringToReturn];
```

Here, `stringToReturn` is the link to our image in the iOS library.

If it fails, we will see the following result:

```
pluginResult = [CDVPluginResult resultWithStatus:CDVCommandStatus_
ERROR];
```

After that, we simply need to invoke the `sendPluginResult` method:

```
[self.commandDelegate sendPluginResult:pluginResult callbackId:self.
callbackID];
```

Here, we passed the `pluginResult` object and `callbackID` that we stored in the beginning.

That is it! Now, we can publish our plugin online.

Publishing and using the plugin

Now, we can publish our plugin to some source control. In my case, I published it to GitHub.

 You can fork the plugin at `https://github.com/cybind/cordova-pugin-imagetolibrary`.

Once it is published, other PhoneGap/Cordova developers can install it by running this command in the console:

```
$ cordova plugin add https://github.com/cybind/cordova-pugin-
imagetolibrary.git
```

This command should be run from the Cordova project root folder:

```
$ cd cordova
$ cordova plugin add https://github.com/cybind/cordova-pugin-
imagetolibrary.git
Fetching plugin "https://github.com/cybind/cordova-pugin-imagetolibrary.
git" via git clone
Repository "https://github.com/cybind/cordova-pugin-imagetolibrary.git"
checked out to git ref "master".
Installing "com.cybind.imaginary.imagetolibrary" for ios
```

We successfully installed the plugin, and we can now use it.

Now, we can define a method in the main controller to save pictures into the library. Let's name it `saveToLibrary`:

```
saveToLibrary: function(base64Data, callback) {
    window.ImageToLibraryPlugin.saveToLibrary([base64Data],
    function(imageUrl) {
        if (callback) callback(imageUrl, null);
```

```
    }, function(error) {
        if (callback) callback();
    });
}
```

Here, `base64Data` is the image representation, and `callback` is the function we call after plugin execution with a link to the image in the library.

Now, let's go back to the `copyPhotoToPersistentStore` function and wrap its content with the `saveToLibrary` method:

```
copyPhotoToPersistentStore: function(fileURI, callback) {
    var self = this;
    self.saveToLibrary(fileURI, false, function(imageUrl) {
        // the rest of saving logic goes here
    });
}
```

We are done with the plugin implementation and setup.

Displaying the list of photos

The last thing we can do is to populate the view with all the captured images.

Let's create the controller in the `app/controller/Pictures.js` file:

```
Ext.define('Imaginary.controller.Pictures', {
    extend: 'Ext.app.Controller',
    config: {
        refs: {
            photoContainer: '#photos'
        }
    },
    launch: function() {
        var pictures = this.getPictures();
        var photoContainer = this.getPhotoContainer();
        for (var i = 0; i < pictures.length; i++) {
            this.addPictureToContainer(pictures[i],
            photoContainer);
        }
    }
    // getPictures and addPictureToContainer implementation goes
    here
});
```

The main purpose of the controller is to retrieve all pictures from the pictures store, create images, and add them to the container, as presented here in the `addPictureToContainer` method:

```
addPictureToContainer: function(picture, container) {
    var self = this;
    var thumb = Ext.create('Ext.Img', {
        src: picture.get('url'),
        height: 80,
        width: '20%',
        border: 3,
        style: 'float: left; border-color: white; border-style:
        solid;'
    });
    thumb.picture = picture;
    container.add(thumb);
}
```

Here, we created the Sencha Touch Image object with the source as a URL from the store. Also, we put the width as 20 percent, so there always will be five elements in the row on any width. In addition we styled the border of each picture a little bit. After that, we assigned the picture object from the store to the custom property on the image. It will help us reference the metadata later.

When we launch the application and go to the **My Photos** tab, we will be able to see something like this:

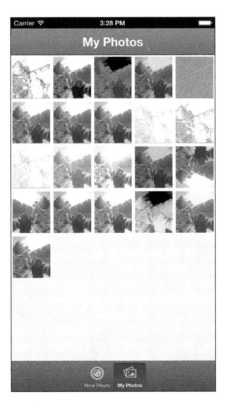

In the preceding screenshot, you are not able to see colored pictures, because the book is printed in black and white. You might be able to see the original picture in the electronic version of the book.

Now, we will implement the picture preview. Let's start by defining the view.

I would like to add the view under app/view/Picture.js. It will be almost the same as a popup preview we have already implemented. We will assign a new name to the view, Imaginary.view.Picture. We only got rid of the effects list and added two other buttons at the bottom: the **Delete** and **Close** buttons:

```
{
    xtype: 'button',
    id: 'deletePhotoBtn',
    text: 'Delete',
    iconCls: 'trash',
    ui: 'decline',
    flext: 1,
    margin: '0 5 0 5'
},
{
    xtype: 'button',
    id: 'closePhotoBtn',
    text: 'Close',
    iconCls: 'delete',
    flext: 1,
    margin: '0 5 0 5'
}
```

The buttons are centered and placed close to each other. Also, I have added some nice icons.

To open the popup, let's attach a tap event handler to the thumbnails:

```
addPictureToContainer: function(picture, container) {
    //...
    thumb.on('tap', function() {
        var thumb = this;
        var popup = Ext.create('Imaginary.view.Picture');
        Ext.Viewport.add(popup);
        popup.show();
        Ext.getCmp('photoPreview').setSrc(thumb.picture.get('url'));

        popup.on('hide', function() {
            popup.destroy();
        });
        Ext.getCmp('deletePhotoBtn').on('tap', function() {
```

```
            // delete picture code goes here
        });
        Ext.getCmp('closePhotoBtn').on('tap', function() {
            popup.hide();
        });
    });
    //...
}
```

As you can see here, we created an instance of the `Imaginary.view.Picture` view, added it to the viewport, and displayed it. After that, we took the source of the thumbnail and assigned it as a source of the image preview in the popup.

Once the popup is hidden, we destroyed it; and once the close button is tapped, we closed the popup.

The delete logic is totally the same as we described in *Chapter 3, Easy Work with Device – Your First PhoneGap Application "Travelly"*.

Now, we can check how the preview works. Run the following command:

```
$ sencha app build -run native
```

Click on the **My Photos** tab, and after that, click on any thumbnail. We will be able to see the popup with the picture and buttons, as shown here:

We are able to see the same picture in the library as well:

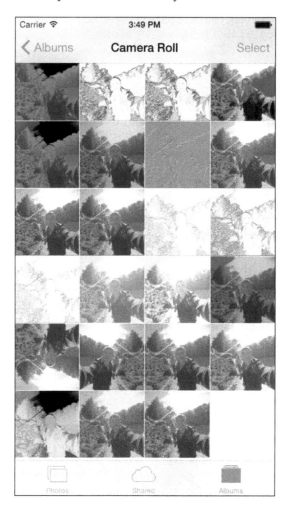

At this stage, we have finished developing the Imaginary application.

Summary

In this chapter, you learned how to work with third-party JavaScript libraries in the Cordova application. We used a camera, filesystem plugins, and several Sencha Touch views and controllers. Also, you learned how to build a Cordova plugin and how to distribute it.

The next chapter is dedicated to common approaches to testing PhoneGap applications. We will implement several UI and unit tests.

Testing the PhoneGap Application

9

In the previous chapter, we wrote an Instagram-like application to apply effects to our pictures with the help of the Pixastic library. Also, we built custom PhoneGap/Cordova plugins for iOS. In this chapter, we will talk about testing, why it is so important, and how to integrate it in our workflow. Also, we will review different types of testing, frameworks and tools to do automated tests. However, before moving forward with the testing description, let's look at the debugging possibilities with PhoneGap.

Running with PhoneGap

There is a PhoneGap Developer App tool available to test our Cordova/PhoneGap applications quickly. Using this tool, we can quickly change code on the desktop and test our application on attached devices without having to rebuild, reinstall, or do other manipulations through an IDE or CLI.

The idea is in connecting the desktop and the mobile device wirelessly over a shared network.

There are two parts in the PhoneGap Developer App tool:

- The mobile application, which is a kind of "shell" for our application
- The mobile PhoneGap application itself, which we are serving over a web server from our computer with the PhoneGap CLI `phonegap server` command

The mobile part will look and act like our mobile installed application. Any edits to the code on the desktop are reflected immediately once we save the code.

Let's look at the details of setup and the usage of the application.

PhoneGap Developer App setup

In all the previous chapters, we had to deal only with Cordova and its CLI. Now, we will use the PhoneGap CLI for different types of running applications on the device. In the *Chapter 1, Installing and Configuring PhoneGap,* we already installed PhoneGap using the following command:

```
$ sudo npm install -g phonegap
```

Now, we can go to any Cordova application created earlier and run the serving command there. Let's try it on the Crazy Bubble application we developed in *Chapter 6, Share Your Crazy Bubbles Game Result on Social Networks*:

```
$ cd crazy-bubbles2
$ phonegap serve
```

After that, we will receive trace messages about the progress and execution process on the server where the application is hosted. Here is an example of how my console output looks:

```
[phonegap] starting app server...
[phonegap] listening on 192.168.0.102:3000
[phonegap]
[phonegap] ctrl-c to stop the server
[phonegap]
```

After that, we will install the mobile part of PhoneGap Developer App on our mobile device.

> You can download the mobile application for iOS, Android, and Windows Phone from http://app.phonegap.com/.

Once it is installed, we can open it and when it prompts, enter the IP address with port in the **Server Address** field. In my case, it is 192.168.0.102:3000. Then, tap **Connect**:

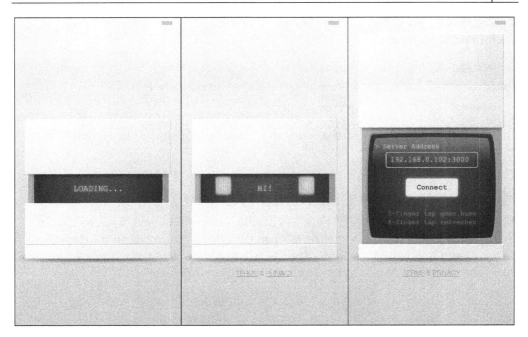

After that, we will see a success message followed by the application running on the device:

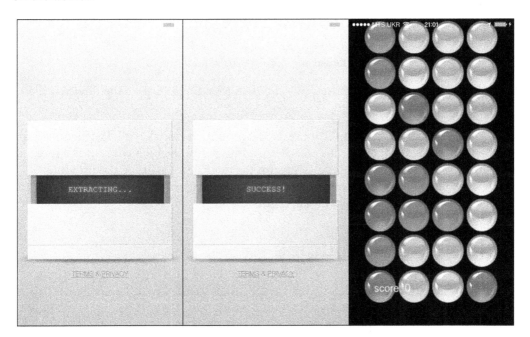

At the same time, on the desktop, in the command line, we will see something like the following command:

```
[phonegap] 200 /__api__/appzip
[phonegap] [console.log] Received Event: deviceready
[phonegap] 200 /__api__/autoreload
```

Here, `Received Event: deviceready` is a `console.log` output in our application in the `app.js` file.

Handling code changes on the fly

Now, we can open our favorite editor to modify our code, save it, and see all our changes immediately on our device. We can do a three-finger tap on the device screen at any time while our application is running to return to the home screen or do a four-finger tap to reload the application with updated content.

Now, let's change the content of a file in the application. I simply commented some style in the CSS file, and here is the output I get in the console:

```
[phonegap] file changed /Users/cybind/Projects/phonegap-by-example/crazy-
bubbles2/www/css/index.css
[phonegap] 200 /__api__/autoreload
[phonegap] 200 /__api__/autoreload
[phonegap] 200 /__api__/appzip
[phonegap] [console.log] Received Event: deviceready
```

If we do not see any files reloading, we can do a four-finger tap on the device screen to reload our application.

We can serve other mobile applications to the mobile device as well. We just need to stop the current serving with *Ctrl* + *c*, then cd into the project we want to run next, and simply type `phonegap serve`. Then, on the device, we will see the previous application still running. We only need to do a four-finger tap, and it will refresh our screen with the new application content.

Including core plugins

The PhoneGap Developer App has already included PhoneGap core plugins in it, so we do not need to add plugins to our application to quickly test some native functions. It helps to eliminate issues when we forget to add some plugins when we develop and debug the application with CLI. However, anyway, we will need to add the plugins before application distribution.

However, there is an issue: `phonegap serve` doesn't support third-party plugins. The reason is that the companion PhoneGap Developer App would need to be compiled with every plugin's native code. Hopefully, it will be improved soon. For now, just JavaScript-only plugins are allowed with this tool.

I highly recommend you to try the PhoneGap Developer App. It is very easy to use.

Why we need tests

We are software developers, and we develop some applications. In this particular case, we develop mobile applications. The code that we write could be put in the browser, as a mobile program, or started as a Node.js script. In different cases, we expect particular results. Every line of code means something, and we need to know that the final product is doing what we need. Normally, we debug our applications. For example, we write part of the application and run it. We use different debugging tools and approaches. Sometimes, we use more complex tools, such as GapDebug, or something simple, such as visual acceptance or `console.log`. By monitoring the output or looking at the way it behaves, we know whether everything is good or there is a problem. However, this approach takes time, especially if the project is big and growing. Iterating over and over again through every single feature of the application could cost a lot of time and money. Automatic testing could help in such cases. Testing is very helpful in better code writing and architecture organization. That's because when the system is complex and when we have a lot of relationships between the modules, it is difficult to add something new or make big changes. We can't guarantee that everything will work as it worked before the modifications. So, instead of relying on our manual testing, it is much better to create scripts that are doing the testing for us. Another name for this process is regression testing. Writing tests have several benefits, which are as follows:

- We save a lot of time, because we don't have to perform manual testing again and again.

- The tests prove that our software is stable and works as expected.

- A badly written code with a lot of dependencies cannot be tested easily. Writing test helps identify problematic parts of code and fix them later.

- If we have a solid test suite, we would be able to extend the system without worrying about breaking something.

- Tests could be used as documentation to the code if the coverage is big enough.

Testing theory

There are few popular ways of writing tests. Let's look at them and understand the difference between them.

Test-driven development

Test-driven development (TDD) is a process that relies on short repetition development cycles. It means that we write tests while we are writing an implementation. The shorter the cycles are, the better. Here is a diagram showing the TDD flow:

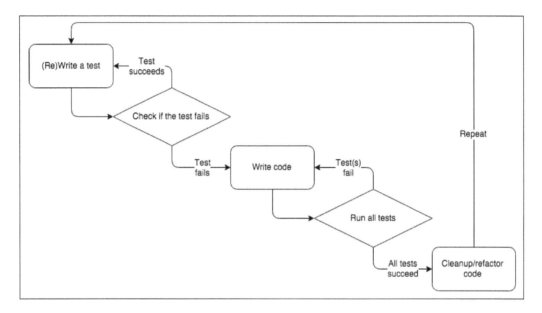

Before we write the actual code that does the job, we need to prepare a test. After the first run, the test fails, because there is nothing implemented. After that, we have to make the test pass cycle by cycle. When it happens, we may spend some time refactoring what is done so far and continue with the next method, class, or feature. Everything starts from writing tests. This is a very good approach, because that's the place where we define what our code should do. By writing tests, we protect ourselves from writing and delivering code that is not needed. We are also sure that the implementation meets the needed requirements.

Let's summarize the steps we need to follow in each iteration for TDD:

1. Add a test.
2. Run these tests and check whether at least one is failing.
3. Write some code.
4. Run tests.
5. Refactor code.
6. Repeat the previous steps.

There is another related concept in testing named — **Acceptance test-driven development (ATDD)**. It is a development methodology based on communication between the customer, developer, and tester. It is placed at a higher level than TDD and includes a variety of other testing approaches. For example, we can review **behavior-driven development (BDD)**.

Behavior-driven development

Behavior-driven development is similar to TDD. If the project is small, we can't tell the difference. The idea of this approach is to focus more on the specification and the processes happening in the application rather than on the actual code. For example, if we test a module that posts messages to our blog with TDD, we would probably ask:

- Is the message empty?
- Is the user authenticated?
- Is the POST request made properly?
- Does the returned page contain posted data?

However, with BDD, we ask:

- Is the message sent to the blog?

Both processes are really close to each other. As we said, in some cases, there is no difference at all. I can tell that the BDD approach is presented at a higher level and could be read by customers. What we should remember here is that BDD focuses on what the code is actually doing and TDD focuses more on how the code is doing it.

Tests classification

There are a lot of different test types. You may already be familiar with some of them, but let's review the most popular ones.

Unit testing

Unit testing performs checks on a single part of the application. It is focused only on one part of the code, one unit. Very often, it is difficult to write such tests because we can't split the code into units. This is usually not good for application scalability and maintenance. If there are no clearly defined modules, we can't proceed with such tests. Having logic put into different units helps not only with the testing but with the whole stability of our program.

Integration testing

Integration tests are those tests whose output is a result of several units or components. The units are typically code modules' separate applications, client and server parts in the network, and so on. Normally, the integration tests use several modules of the system and guarantee that they work properly together.

Functional testing

Functional tests are very close to integration tests. They are focused on specific functionality in the system, on the output that is as per requirement or not. It may involve several modules or components.

System testing

System tests put our program into different environments. In the context of PhoneGap, this could be running our scripts on different mobile platforms and monitoring the output. Sometimes, there are differences, and if we want to globally distribute our work, we should know that it works on the most popular platforms.

Performance or stress testing

These tests put our application beyond the defined specifications and show results of how our code reacts to heavy traffic or complex queries. It could be heavy load similar to putting a large amount of data beyond storage capacity, a large number of concurrent connections, and so on. They are really helpful when we need to take a decision about the program's architecture or choose a framework. There are many different tools to do performance and load testing.

Unit testing frameworks and test runners

There are a lot of testing frameworks nowadays. In general, when we use a framework for testing, we need two things:

- **Test runner**: This is the part of the framework that runs our tests and displays messages telling us whether they have passed or failed.

- **Assertions**: These methods are used for the actual checks. For example, if you need to see whether an active variable is true, then you may write `expect(active).toBe(true)` instead of just `if(active === true)`. It's just better for the reader of the test, and it also prevents some strange situations, for example, if you want to see whether a variable is defined. The `if` statement in the following code returns `true` because the status variable has a value, and this value is `null`. In fact, we are asking "is the status variable initialized", and if we leave the test like that, we will get wrong results. That's why, we need an assertion library that has proper methods for testing, as shown in the following code:

```
var status = null;
if(typeof status != "undefined") {
    console.log("status is defined");
} else {
    console.log("status is not defined");
}
```

The common testing dependency looks like this:

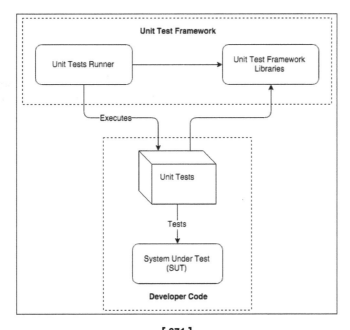

Here are some popular JavaScript testing frameworks:

- **QUnit** is a jQuery test harness
- **Mocha** is a complete JavaScript test framework
- **JSSpec** is a behavior-driven development framework
- **Jasmine** is a behavior-driven development framework; it allows us to do DOM-less testing and asynchronous testing

There are many other JavaScript testing frameworks. Let's look at the Jasmine framework as a testing framework for the Travelly application.

Testing with Jasmine and headless browser PhantomJS

In the following section, we will look at BDD tests implementation with Jasmine and PhantomJS, using the example of the Imaginary application we built in *Chapter 8, Building "Imaginary" – An Application with Instagram-like Image Filters*.

Introduction to the Jasmine

Jasmine is available as a Node.js module and also as a library that we could use in the browser. It comes with its assertion methods.

It's a module, so it could be installed via the Node.js package manager, npm:

```
$ npm install jasmine-node -g
```

The command sets up Jasmine globally, so we can run jasmine-node in every directory of our choice. The tests could be organized into different files placed in one folder or subfolders. The only requirement is to end our file names with spec.js, for example, testing-picture-capture.spec.js or testing-picture-list.spec.js.

Let's look at Jasmine on a particular example.

Writing unit tests with Jasmine

Let's imagine that we want to build an application that reads file content and finds the specific word `dolor`. In the beginning, we will create a project structure:

```
.
├── app.js
├── intro.txt
└── tests
    └── test.spec.js
```

We will place the code that tests the application in `tests/test.spec.js`. The implementation of the logic will be in `app.js`, and the file that we will read from is `intro.txt`. Let's open the file and add some **lorem ipsum** text inside.

The task is pretty simple and can be solved easily in about 10 rows of code. However, we will follow the TDD approach, and separate the code into two different functions, and write tests for each of them:

- Reading the file's content
- Searching for a certain word inside the file's content

We will write tests first, check how they fail, write code to do a job, and test again.

Every test written with Jasmine starts with the `describe` clause. We will start with the following test code:

```js
describe("Testing the reading of the file's content.", function() {
    it("should require of app.js", function(done) {
        var app = require("../app.js");
        expect(app).toBeDefined();
        done();
    });
    it("should read the file", function(done) {
        var app = require("../app.js");
        var content = app.read("./intro.txt");
        expect(content).toBe("Lorem ipsum dolor sit amet, diceret
        percipit ea cum, porro vituperata comprehensam usu ea,
        tation fastidii molestiae eu nam.");
        done();
    });
});
```

Here, the described method accepts a description and a function. In the body of that function, we add our assertions. Similarly, to `describe`, we add `it` blocks. There is a testing logic inside `it` blocks.

As a first argument, we added meaningful information telling what exactly we are going to test. The second argument of it is again a function. The difference is that it accepts an argument, which is another function. We need to call `it` once we finish with the checks. Many things in JavaScript are asynchronous, and the `done` callback helps in handling such operations.

The preceding block requires the `app.js` module and verifies the result. The `expect` method accepts a subject of `assertion`, and the chained methods that follow perform the actual check.

In the second `it` block, we try to read file content and verify that there is proper text inside the file.

Now, we can run the test and see the result:

```
$ jasmine-node ./tests
.F


Failures:

  1) Testing the reading of the file's content. should read the file
   Message:
     TypeError: undefined is not a function
   Stacktrace:
     TypeError: undefined is not a function
    at null.<anonymous> (/Users/cybind/Projects/phonegap-by-
    example/testing/jasmine/intro/tests/test.spec.js:9:21)
    at null.<anonymous> (/usr/local/lib/node_modules/jasmine-
    node/lib/jasmine-node/async-callback.js:45:37)


Finished in 0.014 seconds
2 tests, 2 assertions, 1 failure, 0 skipped
```

The first `it` block goes well, but the second one raises an error. This is because there is nothing in `app.js`. We don't have a `read` method there. Let's add some code to `app.js`, as follows:

```
module.exports = {
    read: function() {
    }
}
```

The test will fail, but due to another reason:

```
$ jasmine-node ./tests
.F


Failures:


  1) Testing the reading of the file's content. should read the file
    Message:
      Expected undefined to be 'Lorem ipsum dolor sit amet, diceret
      percipit ea cum, porro vituperata comprehensam usu ea, tation
      fastidii molestiae eu nam.'.
    Stacktrace:
      Error: Expected undefined to be 'Lorem ipsum dolor sit amet,
      diceret percipit ea cum, porro vituperata comprehensam usu ea,
      tation fastidii molestiae eu nam.'.
     at null.<anonymous> (/Users/cybind/Projects/phonegap-by-
     example/testing/jasmine/intro/tests/test.spec.js:10:19)
     at null.<anonymous> (/usr/local/lib/node_modules/jasmine-
     node/lib/jasmine-node/async-callback.js:45:37)


Finished in 0.013 seconds

2 tests, 2 assertions, 1 failure, 0 skipped
```

This time, the `read` method is defined, but it doesn't return anything, and the `expect` function fails. Now, let's add code to read the file content:

```
var fs = require('fs');
module.exports = {
    read: function(filePath) {
        return fs.readFileSync(filePath).toString();
    }
}
```

Now, the color of the test is in green, indicating that the module has the method that we are trying to use. This method returns what we expect:

```
$ jasmine-node ./tests

..


Finished in 0.016 seconds

2 tests, 2 assertions, 0 failures, 0 skipped
```

To test whether we can find the proper word in the text, we will use the following code:

```
describe("Testing if the file contains certain words", function() {
    it("should contains 'dolor'", function(done) {
        var app = require("../app.js");
        var found = app.check("dolor", "Lorem ipsum dolor sit
        amet");
        expect(found).toBe(true);
        done();
    });
});
```

We require a check method that accepts two arguments. The first one is the word we want to find, and the second one is the string that should contain it.

Just add the testing code to the bottom of test.spec.js and run the test again. It will fail with this message:

```
1) Testing if the file contains certain words should contains
'dolor'

 Message:
   TypeError: undefined is not a function
```

We will go back to our app.js file and add the check function to it:

```
var fs = require('fs');
module.exports = {
    //...
    check: function(word, content) {
        return content.indexOf(word) >= 0 ? true : false;
    }
}
```

The test now passes saying that there are three assertions:

```
$ jasmine-node ./tests

...


Finished in 0.01 seconds

3 tests, 3 assertions, 0 failures, 0 skipped
```

Writing an integration test with Jasmine

Now, we can write a simple integration test that will test how our units work together. We will add the following block to our test file:

```
describe("Testing the whole module", function() {
  it("read the file and search for 'dolor'", function(done) {
    var app = require("../app.js");
    app.read("./intro.txt")
    expect(app.check("dolor")).toBe(true);
    done();
  });
});
```

 Notice that we are not keeping the content of the file in a temporary variable, and we are not passing it to the check method. In fact, we are not interested in the actual content of the file. We are interested only if it contains the specific string. So, our module should handle that and keep the text in it.

So, once we modify our code properly, as shown here, the test will pass:

```
var fs = require('fs');
module.exports = {
  fileContent: '',
  read: function(filePath) {
    var content = fs.readFileSync(filePath).toString();
    this.fileContent = content;
    return content;
  },
  check: function(word, content) {
    content = content || this.fileContent;
    return content.indexOf(word) >= 0 ? true : false;
  }
}
```

That is pretty much how Jasmine works. Let's look into it in real application we already built. I will use the Imaginary application.

Writing Jasmine tests for Sencha Touch's Imaginary application

First of all, we need to download and extract the standalone Jasmine package. We can download it from `http://jasmine.github.io/` and unzip it in the `spec/lib/jasmine` folder in the Imaginary project. Let's create the following structure of the folders:

```
.
├── helpers
│   └── SpecHelper.js
├── javascripts
│   ├── controller
│   ├── model
│   ├── sanitySpec.js
│   ├── store
│   └── view
├── lib
│   └── jasmine
└── spec-runner.html
```

Where:

- `spec-runner.html` is out the main HTML file, where we included Jasmine itself, all testable files, vendor libraries, tests, and test helpers

- `helpers` is a folder where we will store Jasmine helpers

- `javascript` is a folder with actual tests

- `lib` is a folder with the Jasmine library and possibly, some other libraries

Once we downloaded and copied Jasmine to the proper place, we can start setting up `spec-runner.html`. The content of the file in the beginning might look like this:

```html
<html>
<head>
    <title>Sencha Unit Test: Imaginary App Jasmine Tests</title>

    <!-- Jasmine -->
    <link rel="shortcut icon" type="image/png"
    href="lib/jasmine/lib/jasmine-2.3.4/jasmine_favicon.png">
    <link rel="stylesheet" type="text/css"
    href="lib/jasmine/lib/jasmine-2.3.4/jasmine.css">
    <script type="text/javascript" src="lib/jasmine/lib/jasmine-
    2.3.4/jasmine.js"></script>
```

```
    <script type="text/javascript" src="lib/jasmine/lib/jasmine-
    2.3.4/jasmine-html.js"></script>
    <script type="text/javascript" src="lib/jasmine/lib/jasmine-
    2.3.4/boot.js"></script>


    <!-- CODE TO TEST -->


    <!-- UNIT TEST CLASSES -->

</head>
<body>
</body>
</html>
```

As you can see, we included all required Jasmine files here.

We have dependency from Sencha Touch in the application, so we need to add it to `spec-runner.html` as well. We can do this easily by adding the following code after Jasmine:

```
<script type="text/javascript" src="../touch/sencha-touch-all.js"> </
script>
```

Yes, we need it to use Sencha's `Ext.Loader` and `Ext.Application` in our tests. What we need to define right after Sencha Touch is a helper. Let's put it in `helpers/SpecHelper.js`:

```
Ext.require('Ext.data.Model');


afterEach(function () {
    Ext.data.Model.cache = {};
});
```

`Ext.data.Model` caches every model created by our application. If we do not clear this cache, we will get ridiculous results in our tests. We need to include this script as well:

```
<script type="text/javascript" src="helpers/SpecHelper.js"> </script>
```

After that, let's implement some sanity test and put it into the `spec/javascript/sanitySpec.js` file:

```
describe("Sanity", function() {
  it("succeeds", function() {
    expect(true).toEqual(true);
  });
});
```

To see the result of this test, let's run Sencha's web server:

```
$ sencha web start
Sencha Cmd v5.1.3.61
[INF] Mapping http://localhost:1841/ to ....
[INF] -------------------------------------------------------------
[INF] Starting web server at : http://localhost:1841
[INF] -------------------------------------------------------------
```

Now, we can open test runner from `http://localhost:1841/spec/spec-runner.html`. In the browser, we will see something like this:

Do not forget to include the test in `spec-runner.html`:

```
<script type="text/javascript" src="javascripts/sanitySpec.js"> </script>
```

Now, we can include the required units to test controllers and models.

Writing Jasmine tests for a controller

Before writing the actual test, we need to include a controller in `spec-runner.html`:

```
<script type="text/javascript" src="../app/controller/Main.js"> </script>
```

Let's place the test in `javascripts/controller/MainControllerSpec.js`:

```javascript
describe('Imaginary.controller.Main', function() {
    var controller, app;
    beforeEach(function() {
        app = Ext.create('Ext.app.Application', {
            name: 'Imaginary'
        });
        controller = Ext.create('Imaginary.controller.Main', {
            application: app
        });
```

```
        controller.launch();
    });

    afterEach(function() {
        app.destroy();
    });

    //...
});
```

Here, we used the `beforeEach` function to define some logic, which should be executed before each test run in this file. Here, we created our application's controller and executed the launch method of the controller. So, by doing this, we mimicked the initial loading steps of the application so that we can use the controller for testing.

Using `afterEach`, we can destroy objects that we do not need after test execution.

In the `Main` controller, we have the `getCameraPicture` method, which should return the picture test URL:

```
Ext.define('Imaginary.controller.Main', {
    extend: 'Ext.app.Controller',

    //...

    getCameraPicture: function(callback) {
        //...
        // Emulate captured image
        if (callback) callback('resources/images/test.jpg');
        //...
    },

    //...

});
```

Let's write a test for this function to check whether we can call the function and whether we get the expected result:

```
describe('#getCameraPicture', function() {

    it('calls to get camera picture', function() {
        spyOn(controller, 'getCameraPicture');
        controller.getCameraPicture();
        expect(controller.getCameraPicture).toHaveBeenCalled();
    })
```

```
it('passes testing url to callback', function() {
    controller.getCameraPicture(function(url) {
        expect(url).toEqual('resources/images/test.jpg');
    });
})

});
```

These tests are pretty simple, so let's just refresh the test runner page in the browser. We will see that these tests have succeeded:

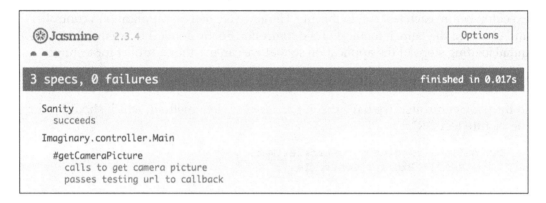

Writing Jasmine tests for a model

Similarly, we will write tests for a Sencha model. Include both model and test for the model:

```
<script type="text/javascript" src="../app/model/Picture.js"> </
script>
<script type="text/javascript" src="javascripts/model/
PictureModelSpec.js"></script>
```

We will do two simple checks now:

- Check whether the model exists
- Check whether the model has data

The content of `spec/javascripts/model/PictureModelSpec.js` looks like this:

```
describe('Imaginary.model.Picture', function() {
    it('exists', function() {
        var model = Ext.create('Imaginary.model.Picture');
        expect(model.$className).toEqual
        ('Imaginary.model.Picture');
    });

    it('has data', function() {
        var model = Ext.create('Imaginary.model.Picture', {
            id: 1,
            url: 'some/test/url.jpg'
        });
        expect(model.get('id')).toEqual(1);
        expect(model.get('url')).toEqual('some/test/url.jpg');
    });
});
```

Here, we tried to instantiate the `Imaginary.model.Picture` class and check whether it equals its name. After that, we tried to create the model with some data and check whether the data is the same that we once retrieved.

Now, we can run all our tests for controller and model together. We will see the following screenshot with all passing tests:

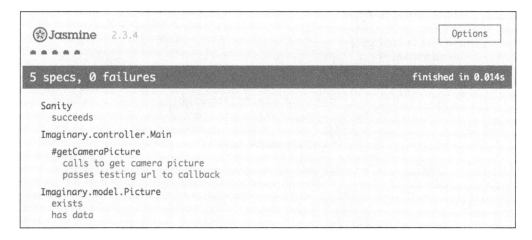

Running tests with the headless browser PhantomJS

Now, we can try to run our tests without a real browser. We can do this with a headless browser. The headless browser that we will use is called **PhantomJS**. It's a browser based on **Webkit** but controlled from the command line or via code. It comes as a binary. Of course, the JavaScript community developed tools to combine testing frameworks such as Jasmine and PhantomJS. We will use one with the name grunt-contrib-jasmine.

Let's install it with npm:

```
$ npm install grunt-contrib-jasmine --save-dev
```

Here, we added the --save-dev option to save the grunt-contrib-jasmine package information in package.json. You can see the grunt prefix in the package name.

 Grunt is a command-line build tool for JavaScript projects. In other words, it is a task runner.

Once we install it, we should add the properly configured Gulpfile.js file. Let's put it in the spec folder:

```
module.exports = function(grunt) {
    'use strict';

    grunt.initConfig({
        jasmine: {
            src: '../app/**/*.js',
            options: {
                specs: 'javascripts/**/*Spec.js',
                helpers: 'helpers/*Helper.js',
                vendor: [
                    '../touch/sencha-touch-all.js'
                ]
            }
        }
    });

    grunt.loadNpmTasks('grunt-contrib-jasmine');
    grunt.registerTask('test', ['jasmine']);
    grunt.registerTask('default', ['test']);

};
```

The main part here is the `jasmine` section in the `grunt.initConfig` function. In this section, we specified the same includes we did in `spec-runner.html`. In this case, we will not use a HTML file and will see all the tests runs in the command line. These are the main options in the `jasmine` section:

- `src` is the path to the testing code
- `specs` is a path to our tests
- `helpers` is a path to the Jasmine test helpers, which should be loaded before tests
- `vendor` is an array with paths to third-party libraries we need to use; in our case, it is Sencha Touch

After that, we register the task with the name `test`, which will execute jasmine commands through PhantomJS. We will make this test the default one. It means we will run the test with the following commands:

```
$ grunt
$ grunt test
```

The result is pretty awesome. We are able to see all the same results in the console now:

```
$ grunt
Running "jasmine:src" (jasmine) task
Testing jasmine specs via PhantomJS


  Imaginary.controller.Main
    #getCameraPicture
      ✓ calls to get camera picture
      ✓ passes testing url to callback
  Imaginary.model.Picture
    ✓ exists
    ✓ has data
  Sanity
    ✓ succeeds


5 specs in 0.011s.
>> 0 failures


Done, without errors.
```

All our tests passed in the browser test runner and in PhantomJS as well.

Now, let's look at testing in a real browser.

Testing with DalekJS in a real browser

All these testing things are really helpful, but it will be even better if we are able to run a real browser and control it. With DalekJS, this is possible. It's really a nice Node.js module that comes with the command-line interface tool and submodules for the major browsers such as Google Chrome, Firefox, and, Internet Explorer.

Let's install the Dalek CLI:

```
$ npm install -g dalek-cli
```

DalekJS supports several browsers including Google Chrome, so we will use it. Of course, we should have it installed onto our system. Once the Dalek CLI is installed, let's create a folder for our test and put package.json there with the following content:

```
{
    "name": "DalekJS-Test",
    "description": "DalekJS Test Description",
    "version": "0.0.1",
    "devDependencies": {
        "dalekjs": "*",
        "dalek-browser-chrome": "*"
    }
}
```

A quick npm install command will create the node_modules directory with both dependencies inside.

DalekJS has a good documentation at http://dalekjs.com/pages/documentation.html.

On this site, we can find methods to load the pages, fill forms, click on buttons, query DOM, and so on.

In our test file, we will add the following testing code:

```
var url = 'http://localhost:1841/';
var title = 'DalekJS test';
module.exports = {
  'should interact with the application': function (test) {
      test
      .open(url)
      .assert.title().is('Imaginary')
      .wait(500)
      .assert.text('#takePhotoBtn', 'Take Photo', 'The title on
      the button "Take Photo"')
      .click('#takePhotoBtn')
      .wait(500)
      .assert.text('#retakePhotoBtn', 'Retake', 'The title on the
      retake button "Retake"')
      .assert.text('#savePhotoBtn', 'Save', 'The title on the save
      button "Save"')
      .assert.text('#closePhotoBtn', 'Close', 'The title on the
      close button "Close"')
      .done()
  }
};
```

Here, we requested `http://localhost:1841/`, which should be running by Sencha web server. Once our test loads the application, we checked whether the title of the page is as expected. After that, we waited for 500 milliseconds to be sure that all our Sencha components are loaded. After timeout, we checked the text on the **Take Photo** button and clicked on it. Then, we waited for the popup to appear. After timeout, we did several UI checks again.

To run the test, we simply need to run the following command in the console:

```
$ dalek test.js -b chrome
```

It will start the Chrome browser and execute all the tests. The test run results should look like this:

```
Running tests
Running Browser: Google Chrome
OS: Mac OS X 10.10.3 x86_64
Browser Version: 43.0.2357.130
```

```
RUNNING TEST - "should interact with the application"
▶ OPEN http://localhost:1841/
✔ TITLE
▶ WAIT 500 ms
✔ TEXT The title on the button "Take Photo"
▶ CLICK #takePhotoBtn
▶ WAIT 500 ms
✔ TEXT The title on the retake button "Retake"
✔ TEXT The title on the save button "Save"
✘ TEXT
0 EXPECTED: Close
0 FOUND: [object Object]
0 MESSAGE: The title on the close button "Close"
✘ TEST - "should interact with the application" FAILED


 4/5 assertions passed. Elapsed Time: 4.01 sec
```

As you can see, we got one assertion failing. This is because we put the wrong selector for the Close button. Let's fix it. Take the following line of code:

```
.assert.text('#closePhotoBtn', 'Close', 'The title on the close button
"Close"')
```

We need to change it to this line of code:

```
.assert.text('#cancelPhotoBtn', 'Close', 'The title on the close
button "Close"')
```

After the test run, we will see that all the tests passed:

```
✔ TEST - "should interact with the application" SUCCEEDED


 5/5 assertions passed. Elapsed Time: 3.46 sec
```

We could even make screenshots of the current browser's screen:

```
var url = 'http://localhost:1841/';
var title = 'DalekJS test';
module.exports = {
  'should interact with the application': function (test) {
    //...
    .screenshot('./screen.jpg')
    .done()
  }
};
```

Here, we took a screenshot of the second screen (popup).

Performance testing with Appium and browser-perf

Performance is a big challenge when you develop for PhoneGap/Cordova. It is always important to look at the code in our application that may make an app junky. There is a tool for developers to help measure frame rates, repaints, layout, and so on. The name of this tool is browser-perf. It is a Node.js-based tool that is taking data from browser developer tools and converting it to performance indicators. It supports iOS and Android Cordova applications.

Let's install it from npm:

```
$ npm install -g browser-perf
```

Before starting testing, we need to install Appium as well.

 Appium is a tool used to automate our application and emulate user interactions such as click, swipe, and so on.

It can be installed through npm as well:

```
$ npm install -g appium
```

 Appium requires installation without using root/Administrator privileges. Otherwise, you will get issues within testing.

Once both tools are installed, we can start writing the test. I will insert the following code into the ios-hybrid-appium.config.json file:

```
{
    "selenium": "http://localhost:4723/wd/hub",
    "browsers": [{
        "platformName": "iOS",
        "platformVersion": "8.3",
        "deviceName": "iPhone 6",
        "app": "/Users/cybind/Projects/phonegap-by-
        example/imaginary/cordova/platforms/ios/build/
        emulator/Imaginary.app"
    }]
}
```

It will connect to the Selenium server at http://localhost:4723/wd/hub that will be started by Appium. Also, we will specify the platform, version, and device name where we would like to run our test. In the app, we will specify the full path to the build for testing.

I guess we are ready for the initial run. Let's start Appium first:

```
$ appium
```

Now, we can execute our performance profiler:

```
$ browser-perf -c ios-hybrid-appium.config.json
```

It will start the iPhone emulator and run the application there, in this case, the Imaginary application. After some time, we will see the testing results in the console output. It will look something like this:

Metrics	Value	Unit	Source
FireAnimationFrame	24.729	ms	TimelineMetr…
FireAnimationFrame_max	0.451	ms	TimelineMetr…
FunctionCall	45.241	ms	TimelineMetr…
FunctionCall_max	5.194	ms	TimelineMetr…
RequestAnimationFrame	135.000	ms	TimelineMetr…
TimerFire_max	0.142	ms	TimelineMetr…
TimerFire	0.714	ms	TimelineMetr…
droppedFrameCount	1	count	RafRendering…
FireAnimationFrame_count	135	count	TimelineMetr…
FunctionCall_count	152	count	TimelineMetr…
TimerFire_count	8	count	TimelineMetr…
numAnimationFrames	130	count	RafRendering…
numFramesSentToScreen	130	count	RafRendering…
FireAnimationFrame_avg	0.183	ms	TimelineMetr…
FunctionCall_avg	0.298	ms	TimelineMetr…
TimerFire_avg	0.089	ms	TimelineMetr…
ExpensiveEventHandlers	0	count	RuntimePerfM…
ExpensivePaints	0	count	RuntimePerfM…
PaintedArea_total	0	sq.pixe…	RuntimePerfM…
Layers	0	count	RuntimePerfM…
NodePerLayout_avg	0	count	RuntimePerfM…
PaintedArea_avg	0	sq.pixe…	RuntimePerfM…
meanFrameTime_raf	16.532	ms	RafRendering…
framesPerSec_raf	60.490	fps	RafRendering…

Here, you can see different metrics for animation duration, frame measurement, and other options.

It is a minimal configuration we can do for initial performance testing. However, there are many other actions we can emulate to observe performance changes.

 You can find other possible actions, metrics, and options on the Wiki GitHub page of the project at `https://github.com/axemclion/browser-perf/wiki/Node-Module---API`.

It is also possible to use GUI for the Appium. You can find it on the official website at `http://appium.io/`. It looks like this screenshot:

You can do everything in the same way you are able to do with the console version. It has **Appium Inspector** inside. By clicking on the magnifying glass icon, we can launch it.

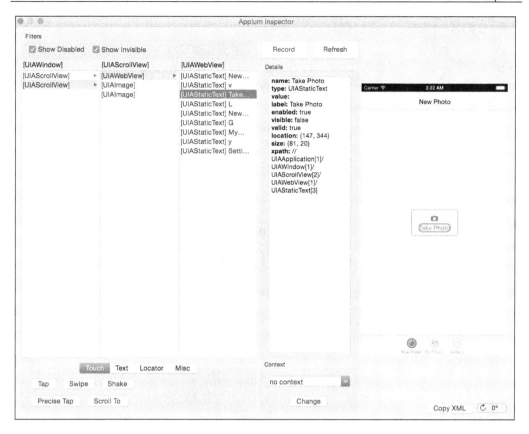

With this tool, it is possible to record tests and execute them later.

I highly recommend you to try Appium and browser-perf for test automation.

Other testing tools

There are many other testing platforms, tools, and services presented on the market. The most interesting of them are services for automated UI testing. Let's review some of them.

Telerik Test Studio

We can use Telerik Test Studio to record and automate UI tests for native, web, or hybrid iOS applications. To create a test, we only need to record our interactions with the application and execute them later.

There are some setup steps we need to follow:

1. Download and install test studio and the mobile testing SDK.
2. Configure a build that can be tested.
3. Deploy the application to the device.
4. Launch test studio, and record a test. You will see help notes on your way.
5. After that, we only need to access the recorded script and execute it.

The only one issue with UI testing is a need to wait until the testing DOM element is rendered. It is possible to set up some timeout or event handler for the UI pre-loader element. It means once the pre-loader is hidden, we can start script execution.

 You can find the product at `http://www.telerik.com/mobile-testing`.

Sauce Labs

This service offers automated cross-browser and mobile testing. It allows us to write Selenium and JavaScript unit tests and run it on many different simulators in the cloud. It is based on Appium.

Also, it allows manual testing on a variety of mobile devices and browsers. So, we can quickly switch between different devices and see how it behaves.

In addition, it is well integrated with **continuous integration** (**CI**) services.

Summary

In this chapter, we saw the importance of testing. Thankfully, there are great tools available in the Node.js ecosystem. Frameworks such as Jasmine make our life easier. Instruments such as PhantomJS save lots of time by automating the testing and addition of our code into a browser. With DalekJS, we were even able to run tests directly in different browsers. Also, we used Appium for cross-platform testing using the same API, which enables code to be reused between iOS and Android test suites.

In the next chapter, we will see how to perform a beta release and deliver it to our testers. Also, you will learn how to distribute our application to the Apple Store and Google Play markets.

10
Releasing and Maintaining the Application

In the previous chapter, we implemented several unit and integration tests with the Jasmine tool for our application. We used the headless browser PhantomJS, and we measured performance with Appium. All this is great and helps us automate the testing approach to find bugs in the early stages of application development. Once we finish creating our application and test it, we can think of delivering our application to other people. We can distribute the application in several different ways.

In this chapter, you will learn how to:

- Build and deploy the application with PhoneGap Build
- Version the application
- Release the beta iOS application with TestFlight
- Release the iOS Cordova/PhoneGap application to the Apple Store
- Release the Android Cordova/PhoneGap application to the Google Play Market
- Distribute and analyze the application with Crashlytics

Once we finish these tasks, we will be ready to do a full cycle of the application creation and distribution processes.

In *Chapter 1*, *Installing and Configuring PhoneGap*, we already discussed how to set up development environments to develop for iOS and Android. We will reuse these skills in this chapter as well to prepare our builds for distribution.

This chapter read as a step-by-step tutorial for the setup of different tools.

Versioning of the application

Versioning is a very important component in application development, maintenance, and support. There are several important cases where we need versioning. These cases are as follows:

- User needs to know specific information about the installed application version and about versions to install for upgrade
- Other installed applications need to know the version of the application to understand their compatibility for interaction
- Different backend services may communicate differently with different application versions

I would recommend that you use the three- number versioning for Cordova/ PhoneGap builds. This versioning is explained here:

- Major version, interface changes, huge application upgrade, and so on
- Minor features, major bug fixes, and so on
- Small mistakes, spelling issues, and so on

There are many different approaches to change these numbers. They could even reflect build dates.

In addition, both iOS and Android support second-version strings:

- `versionCode` for Android
- `CFBundleVersion` for iOS

  Here is an example:

  ```
  <widget id="com.cybind.crazybubbles"
  version="0.0.1"
  versionCode="1"
  CFBundleVersion="0.0.1">
  ```

Now that we have defined our versioning approach, we can start releasing the application through different methods.

Using PhoneGap Build

In the *Chapter 1, Installing and Configuring PhoneGap*, we learned how to build our application using IDE (Xcode or Android Studio). However, now, we will explore how to build the application for different platforms using the PhoneGap Build service.

 PhoneGap Build helps us stay away from different SDKs. It works for us by compiling in the cloud.

First of all, we should register on `https://build.phonegap.com`. It is pretty straightforward. Once we register, we can log in, and under the apps menu section, we will see something like this:

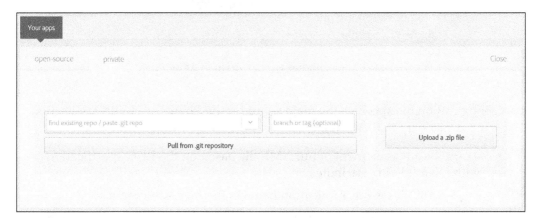

We entered a link to our `git` repository with source files or upload the zip archive with the same source code.

However, there is a specific requirement for the structure of the folders for upload. We should take only the `www` directory of the Cordova/PhoneGap application, add `config.xml` inside it, and compress this folder. Let's look at this approach using an example of the Crazy Bubbles application.

PhoneGap config.xml

In the root folder of the game, we will place the following `config.xml` file:

```xml
<?xml version="1.0" encoding="UTF-8" ?>
    <widget xmlns     = "http://www.w3.org/ns/widgets"
        xmlns:gap     = "http://phonegap.com/ns/1.0"
        id            = "com.cybind.crazybubbles"
        versionCode = "10"
        version       = "1.0.0" >
    <name>Crazy Bubbles</name>
    <description>
        Nice PhoneGap game
    </description>
    <author href="https://build.phonegap.com"
    email="support@phonegap.com">
        Andrew Kovalenko
    </author>
    <gap:plugin name="com.phonegap.plugin.statusbar" />
</widget>
```

This configuration file specifies the main setup for the PhoneGap Build application. The setup is made up of these elements:

- `widget` is a root element of our XML file based on the W3C specification, with the following attributes:
    - `id`: This is the application name in the reverse-domain style
    - `version`: This is the version of the application in numbers format
    - `versionCode`: This is optional and used only for Android

- `name` of the application
- `description` of the application
- `name` of the author with website link and e-mail
- List of plugins if required by the application

We can use this XML file or enter the same information using a web interface. When we go to **Settings** | **Configuration**, we will see something like the following screenshot:

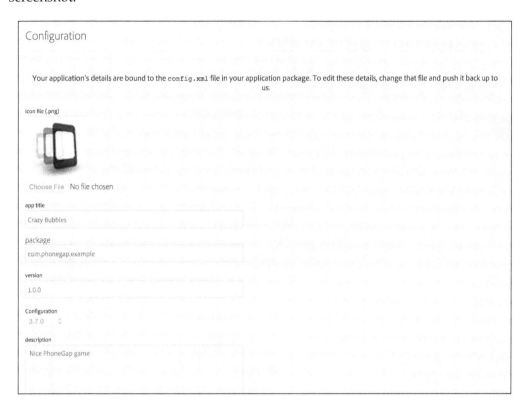

PhoneGap plugins

As you can see, we included one plugin in `config.xml`:

```
<gap:plugin name="com.phonegap.plugin.statusbar" />
```

There are several attributes that the `gap:plugin` tag has. They are as follows:

- `name`: This is required, plugin ID in the reverse-domain format
- `version`: This is optional, plugin version
- `source`: This is optional, can be `pgb`, `npm`, or `plugins.cordova.io`. The default is `pgb`
- `params`: This is optional, configuration for plugin if needed

We included the StatusBar plugin, which doesn't require JavaScript code. However, there are some other plugins that need JavaScript in the `index.html` file. So, we should not forget to add the code.

Initial upload and build

Once we finish the configuration steps and create a Zip archive of the www folder, we can upload it. Then, we will see the following screen:

Here, we can see generic information about the application, where we can enable remote debugging with Weinre.

 Weinre is a remote web inspector. It allows access to the DOM and JavaScript.

Now, we can click on the **Ready to build** button, and it will trigger the build for us.

Here, you can see that the iOS build has failed. Let's click on the application title and figure out what is going on. Once the application properties page loads, we will see the following screenshot:

When we click on the **Error** button, we will see the reason why it failed:

So, we need to provide a signing key. Basically, you need a provisioning profile and certificate needed to build the application. We already downloaded the provisioning profile from the Apple Development portal, but we should export the certificate from the Keychain Access.

We are going to open it, find our certificate in the list, and export it:

When we export it, we will be asked for the destination to store the `.p12` file:

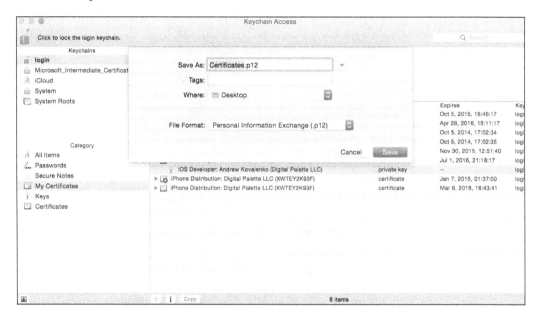

Add a password to protect the file:

Once we save the file, we can go back to the PhoneGap Build portal and create a signing key:

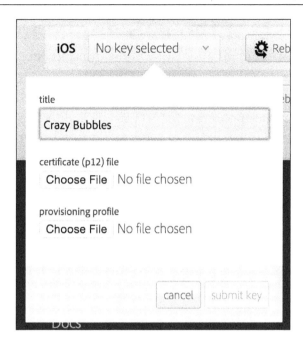

Just click on the **No key selected** button in the dropdown and upload the exported certificate and provisioning profile for the application. Once the upload is finished, the build will be triggered:

Now, we will get a successful result and can see all the build platforms:

Now, we can download the application for both iOS and Android and install it on the device. Alternatively, we can install the application by scanning the QR code on the application main page. We can do this with any mobile QR scanner application on our device. It will return a direct link for the build download for a specific platform. Once it is downloaded, we can install it and see it running on our device.

Congratulations! We just successfully created the build with the PhoneGap Build service! Now, let's take a closer look at the versioning approach for the application.

Beta release of the iOS application

For the beta release of our application, we will use the TestFlight service from the Apple.

As a developer, we need to be a member of the iOS Developer program. As a tester, we will need to install the application for beta testing and the TestFlight application from the App Store. After that, the tester can leave feedback about the application.

First of all, let's go to `https://itunesconnect.apple.com` and login there. After that, we can go to the **My Apps** section and click on the plus sign in the top-left corner. We will get a popup with a request to enter some main information about the application. Let's add the information about our application so that it looks like this:

All the fields in the preceding screenshot are well known and do not require additional explanation.

Once we click on the **Create** button, the application is created, and we can see the **Versions** tab of the application. Now, we need to build and upload our application. We can do this in two ways:

- Using Xcode
- Using Application Loader

However, before submitting to beta testing, we need to generate a provisioning profile for distribution. Let's do it on the Developer portal.

Generate a distribution provisioning profile

Go to the **Provisioning Profiles**, and perform the following steps:

1. Click on **+** to add a new provisioning profile and go to **Distribution | App Store** as presented in the following screenshot:

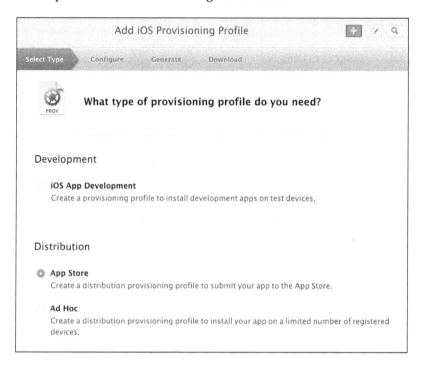

2. Then, select the application ID. In my case, it is `Travelly`:

3. After that, select the certificates to include in the provisioning profile. The certificate should be for distribution as well:

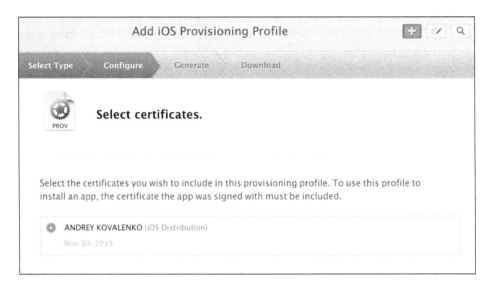

4. Finally, generate the provisioning profile, set a name for the file, and download it:

5. Now, we can build and upload our application to iTunes Connect.

Upload to iTunes Connect with Xcode

Let's open the Travelly application in Xcode. Go to `cordova/platforms/ios` and open `Travelly.xcodeproj`. After that, we have to select **iOS Device** to run our application. In this case, we will see the **Archive** option available. It would not be available if the emulator option is selected. Now, we can initiate archiving by going to **Product | Archive**:

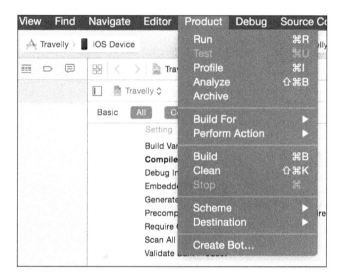

Once the build is completed, we will see the list of archives:

Now, click on the **Submit to App Store...** button. It will ask us to select a development team if we have several teams:

At this stage, Xcode is looking for the provisioning profile we generated earlier. We would be notified if there is no distribution provisioning profile for our application.

Once we click on **Choose**, we are redirected to the screen with binary and provisioning information:

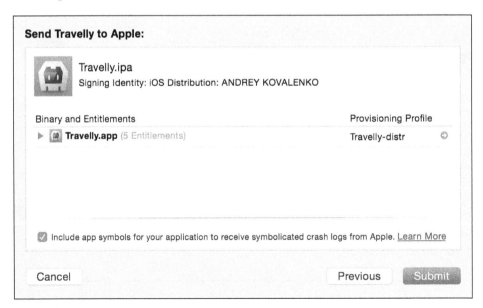

When we click on the **Submit** button, Xcode starts to upload the application to iTunes Connect:

Congratulations! We have successfully uploaded our build with Xcode:

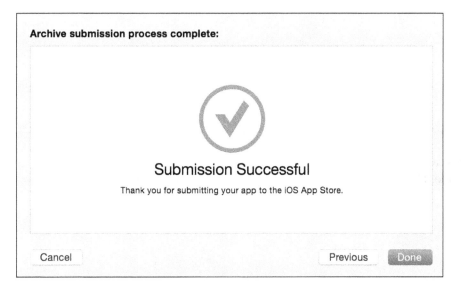

Upload to iTunes Connect with Application Loader

Before the reviewing process of build upload with Application Loader, we need to install the tool first.

Let's go to **iTunes Connect | Resources and Help | App Preparation and Delivery** and click on the **Application Loader** link. It will propose the installation file for download. We will download and install it. After that, we can review the upload process.

Uploading with Application Loader is a little different than with Xcode. We will follow the initial steps until we get the following screen:

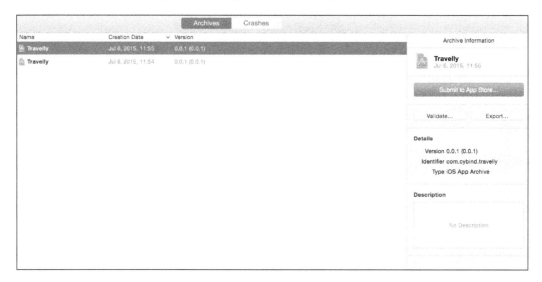

In this case, on the screen, we will click on the **Export** button, where we can save the `.ipa` file. However, before that, we have to select the export method:

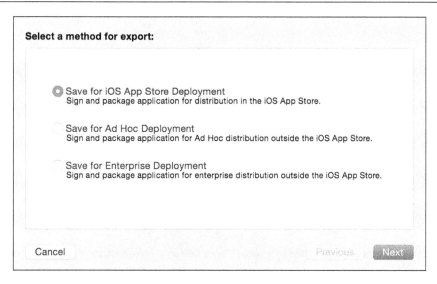

Select a method for export:

○ Save for iOS App Store Deployment
Sign and package application for distribution in the iOS App Store.

Save for Ad Hoc Deployment
Sign and package application for Ad Hoc distribution outside the iOS App Store.

Save for Enterprise Deployment
Sign and package application for enterprise distribution outside the iOS App Store.

Cancel   Previous   Next

We are interested in distribution to the App Store, so we selected the first option. We need to save the generated file somewhere to the filesystem.

Now, we will launch Application Loader and log in using our Apple Developer account:

After that, we will select **Deliver Your App** and pick the generated file:

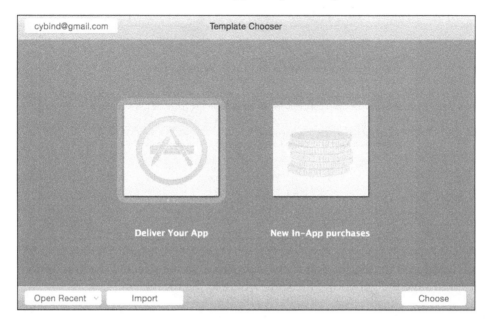

In the following screenshot, we can see the application's generic information: name, version, and so on:

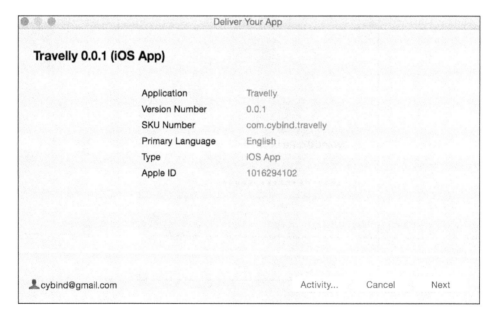

When we click on the **Next** button, we will trigger upload to iTunes Connect, which is successfully executed:

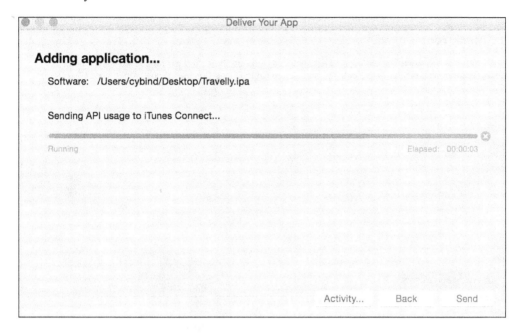

During the process, the package will be uploaded to the iTunes Store, as shown here:

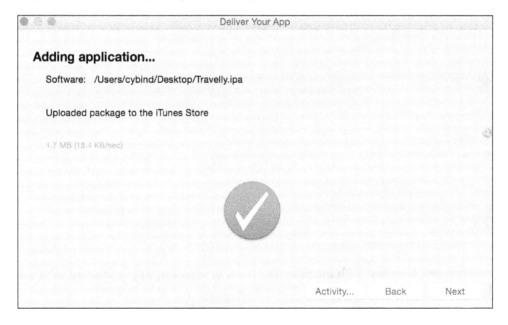

Once the application is added, it will show you the following screenshot:

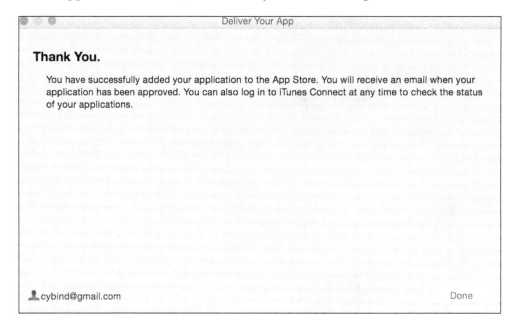

Now, if we go to **iTunes Connect** | **My Apps** | **Travelly** | **Prerelease** | **Builds**, we will see our two uploaded builds:

As you can see, they are both inactive. We need to send our application to internal and external testers.

Invite internal and external testers

Let's work with version 0.0.2 of the application. First of all, we need to turn on the check box to the right of the **TestFlight Beta Testing** label.

There are two types of testers we can invite:

- Internal testers are iTunes Connect users. It is possible to invite up to 25 internal testers.
- External testers are independent users who can install the application using the TestFlight mobile tool.

To invite internal testers, let's go to the **Internal Testers** tab, add the e-mail of the desired tester, place the check mark, and click on the **Invite** button:

The user will receive an e-mail with the following content:

Users can click on the link and follow the instructions to install the application.

To allow testing for external users, we will go to the **External Testers** tab. Before becoming available for external testing, the application should be reviewed. For the review, some generic information is needed. We need to add:

- Instructions for the testers on what to test
- Description of the application
- Feedback information

Once this information is entered, we can click on the **Next** button and answer questions about cryptography usage in the application:

We do not use cryptography, so we select **No** and click on **Submit**. Now, our application is waiting for review approval:

Now, there is a button available to add external testers:

We can invite up to 1000 external testers. After the tester accepts the invite on their device, the invite will be linked to their current Apple ID.

Once the application review is finished, it will become available for external testers.

Release to the App Store

We already uploaded the build for the beta releasing and testing. Releasing to the App Store is almost the same, except that there are requirements in other additional fields for distribution, which are as follows:

- At least one screenshot of the application
- A description of our application
- Keywords that describe our application, which will be used for search in the App Store
- A URL with support information for our application
- Application icon in the .jpg or .png format with a minimum resolution of 72 dpi without rounded corners

- Name of the person who owns exclusive rights to the application
- Category of the application
- Application build that we already uploaded
- Application review information that the application team can use to contact

In addition, we have to add our pricing information:

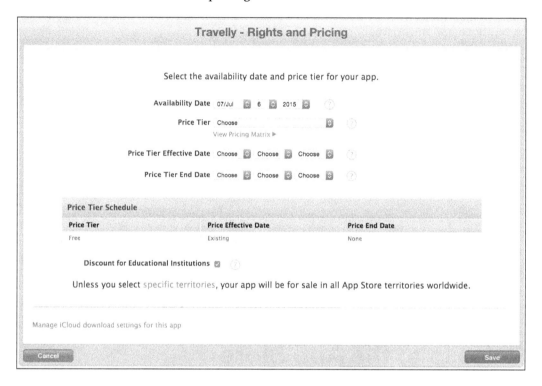

Here, we enter the date when the application will be available, its cost, and some discount information.

Also, we can select a specific build in the **Build** section to deliver with this release:

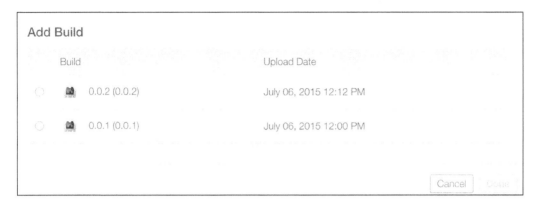

When everything is entered properly and there are no validation errors, we will be able to submit our application for review. You can see the appropriate button in the top-right corner:

Another important item to release to the App Store is the version release approach. It could be:

- **Manual**: We have to manually release the application once it is approved
- **Auto**: The application will be released automatically once approved

And that is it! If we do not break Apple's rules for application development and releasing, then it will be approved in several days. It may vary from two to seven working days.

Release to Google Play

Releasing to Google Play is a little different. First of all, let's check the `cordova/platforms/android/AndroidManifest.xml` configuration. We need to specify:

- `versionCode`: Google Play will not accept the application if `versionCode` is different from the previous versions in the store
- `versionName`: This is not used for anything except to display to users. It's a string, so we can name it the way we want
- Set debuggable to `false`, for example, `android:debuggable="false"`

Create a keystore file

Now, we need to create a keystore file and set a password. We should ensure that we do not lose this file because we need it with every submission to the store. If we create a new one, the application would be interpreted as a new application and not as another version of the same application.

 By default, the debug configuration uses the debug keystore located in `$HOME/.android/debug.keystore`.

We can generate a private key using Android SDK and JDK commands. Let's do this with the following command:

```
$ keytool -genkey -v -keystore travelly-release-key.keystore -alias
travelly -keyalg RSA -keysize 2048 -validity 10000
```

It prompts us for a password for the keystore and key and also for distinguished name fields for the key. It generates the `travelly-release-key.keystore` file. This keystore contains a key that is valid for 10000 days. *Alias* is the name we use to sign our application.

Build and sign an application in the release mode

Now, we need to tell ANT where the keystore file is located. We need to create the `platforms/android/ant.properties` file with the following content:

```
key.store=/Users/cybind/.android/travelly-release-key.keystore
key.alias=Travelly
```

To build the application, let's navigate to the `cordova` folder and run the following well-known command:

```
$ cordova build android
```

It will create the following files in the `platforms/android/ant-build` folder:

```
CordovaApp-debug-unaligned.apk
CordovaApp-debug-unaligned.apk.d
CordovaApp-debug.apk
CordovaApp.ap_
CordovaApp.ap_.d
```

Then, we will navigate to the `android` directory and run the `ant release` command:

```
$ cd platforms/android
$ ant release
```

It will prompt us for a password for the keystore and a password for the alias `travelly`.

Once the process is finished, we will see the following message:

```
[echo] Signing final apk...

[zipalign] Running zip align on final apk...

[echo] Release Package: /Users/cybind/Projects/phonegap-by-example/
sencha-travelly/cordova/platforms/android/bin/CordovaApp-release.apk
```

As you can see, in `platforms/android/bin`, we have release versions of the application now:

```
CordovaApp-release-unaligned.apk
CordovaApp-release-unsigned.apk
CordovaApp-release-unsigned.apk.d
CordovaApp-release.apk
```

Now, we can verify our package with the `zipalign` tool:

```
$ zipalign -v 4 CordovaApp-release.apk Travelly.apk
```

After some time, it should show us the **Verification successful** message. It will generate the `Travelly.apk` file in the application folder.

Now, we can submit our application to the Google Play market.

Upload the application to the Google Play market

To submit the application, we need to go to `https://play.google.com/apps/publish`. Of course, we need to create a Developer account. It costs $25 USD.

We need to follow these steps to submit the application:

- Upload `.apk` file (required)
- Enter the following Store Listing information (required):
  - Title
  - Short description
  - Full description
  - Screenshots, icons, and promo videos
  - Category
  - Contact details
  - Link to privacy policy
- Content Rating (required) where we will pass through a questionnaire. It is available once the application is uploaded
- Pricing and Distribution (required) here we would need to set up a merchant account if we are going to sell our application
- In-application purchases
- Interaction with other services and APIs

I will not describe these steps in detail. There is a lot of information related to this online.

Once you enter all the required information, a **Publish app** button will become available. We can click on this button and submit our application to the Google Play:

Using Fabric and Crashlytics

Nowadays, in the world where mobile devices have become a trending platform and form-factor, we need focus on analytics and crash reporting for the mobile applications.

We can use the Fabric tool for these needs.

Fabric is a modular SDK platform that allows easier builds of mobile applications. It includes:

- Integration into the Xcode workspace
- Xcode launcher
- Crashlytics — crash reporting, beta distribution, and mobile analytics
- Twitter kit

Crashlytics is a really lightweight crash-reporting solution. Let's look at how we can use it on a Cordova iOS application example. I will take the Imaginary application in this case.

After registration on the Fabric site (`https://fabric.io`), we will be requested to download and install the Fabric plugin:

We will select the Xcode option here and click on the **Download** button in the following screen:

Once it is downloaded, we can install and launch it. After that, we can log in with the registered Fabric account, and we will see that it has grabbed all the projects we have in Xcode. We will select the Imaginary project and follow the instructions we get after that:

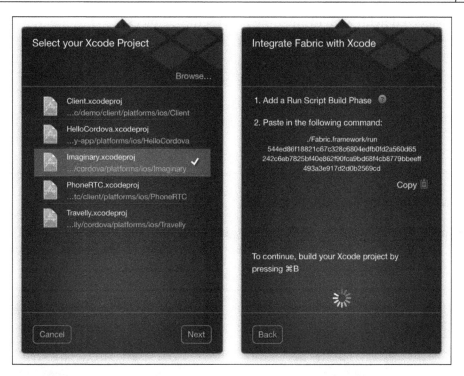

We will open the Imaginary Xcode project, go to **Build Phases**, and add **Run Script Phase**:

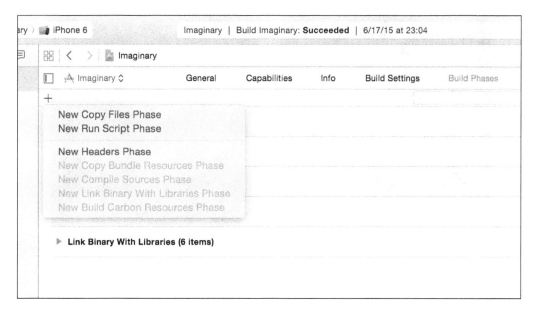

In the displayed field, we will enter the proposed code block:

Then, we will trigger the built-in Xcode.

After that, we will install the SDK by dragging and dropping the icon into the Xcode Project Navigator. Now, we will see instructions on adding code blocks into our application:

We will enter these code blocks in the `Classes/AppDelegate.m` file in the header:

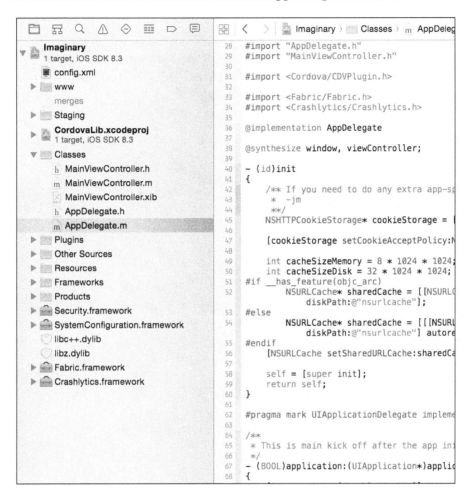

The header files are before the `application` method:

```
64    /**
65     * This is main kick off after the app inits, the views and Settings are
66     */
67    - (BOOL)application:(UIApplication*)application didFinishLaunchingWithOp
68    {
69        [Fabric with:@[CrashlyticsKit]];
70
71        CGRect screenBounds = [[UIScreen mainScreen] bounds];
72
```

Now, we can build our application in Xcode. Once we finish creating the build, we will see the release in Crashlytics:

Now, we can distribute it to our internal beta testers, as we did earlier in the chapter. However, in this case, we can do distribute our application with the help of Crashlytics. We should ensure that the provisioning profile for the application includes all devices on which we want to run our application.

When we click on the **Distribute** button, we can enter the testers' names and release description, and initiate the upload.

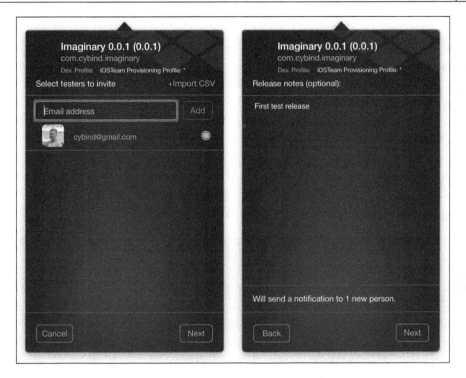

Once the build is uploaded, it will send an invitation to testers. Testers will see e-mails with links to download and install the application. However, before that, it will ask you to set up the Crashlytics profile:

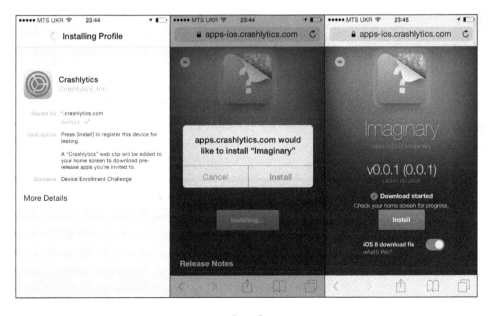

Now, we have successfully installed our application on the mobile device.

If we go to `https://fabric.io` with the registered account, we will be able to see different analytics for the application:

- Users
- New users
- Sessions
- Reports grouped by beta testers
- Crash reports

Here are several screens with analytics and crash reporting:

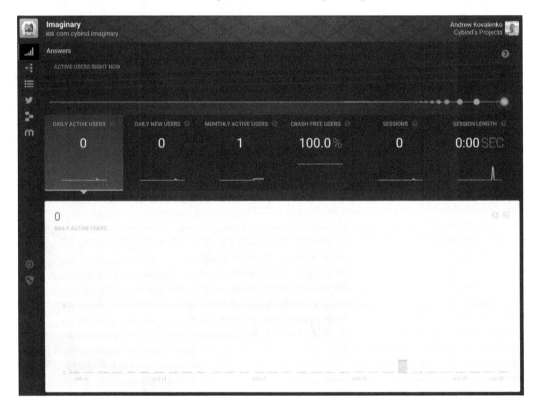

The same crash e-mail notification is seen in the following screenshot:

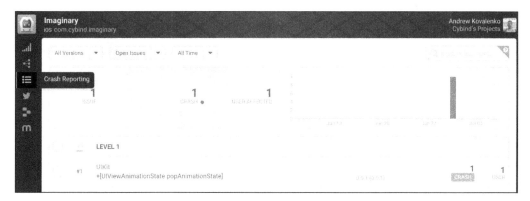

Once we get the crash report in the application and once the administrator gets an e-mail notification as well, he can quickly react to fix the issue.

Summary

In this final chapter of the book, you learned how to release the PhoneGap application with the PhoneGap Build service. Also, we released the application through TestFlight for beta testing. Eventually, you learned how to publish your application to the Apple Store and Google Play markets. This process is not so complicated as it seems the first time. Also, we used Crashlytics to crash report our application. It helped us catch errors and crashes on testers devices so that we are able to maintain it in time.

Now, we will be able to develop different types of Cordova/PhoneGap applications, test them, and deliver them to the markets. I think it is pretty awesome, don't you?

References

You can find all the applications developed in the book on GitHub:

- `https://github.com/cybind/travelly`
- `https://github.com/cybind/travelly-svc`
- `https://github.com/cybind/crazy-bubbles`
- `https://github.com/cybind/pumpidu`
- `https://github.com/cybind/pumpidu-peerjs`
- `https://github.com/cybind/imaginary`
- `https://github.com/cybind/cordova-pugin-imagetolibrary`

Index

A

acceptance test-driven development (ATDD) 269
Android
  Android emulator, adding 34-36
  Android Studio, installation 30-32
  JDK, installation 28
  project, opening in Android Studio 33, 34
  SDK, installation 28-30
  setting up, for application 27
  URL, for Android designs 39
Android debugging setup, GapDebug
  about 71
  Android device configuration 72-75
  computer configuration 72
Android Studio
  URL 30
Apache
  URL 152
Apache Cordova. *See* **PhoneGap**
app directory
  controller 50
  model 50
  profile 50
  store 50
  view 50
Appium
  URL 292
  used, for performance testing 289-293
application
  creating 10, 11
  generating 49-54
  iOS, setting up 14, 15
  releasing, to App Store 319-321
  releasing, to Google Play 322
  running, in iOS emulator 16, 17
  uploading, in Google Play 324
Application Loader
  iOS application, uploading to iTunes Connect 312-316
application, running on iOS device
  application identifier, adding 20, 21
  device, registering 22, 23
  iOS developer certificate, generating 18-20
  performing 17
  Provisioning Profile, generating 24-26
application structure
  about 54-56
  controller, creating 58, 59
  device profiles, creating 64
  environment detection 64
  launch process 65, 66
  model 61, 62
  proxy 64
  store, creating 62, 63
  store, using 60, 61
  theming 66, 67
  UI 66, 67
  view 56-58
App Store
  application, releasing 319-321

B

Base64 87
behavior-driven development (BDD) 269
browser-perf
  used, for performance testing 289-293
browser storage support
  URL 61

bubble, Crazy Bubbles game
  board, refilling 170, 171
  drop down bubbles, using 169, 170
  matched bubbles, removing 169
  matching 166-168
  releasing 166

C

camera
  new picture popup, creating 88-93
  plugin, installing 85
  plugin, using 86-88
  used, for capturing pictures 85
Certificate Signing Request (CSR) 18
Cloud9 IDE
  URL 152
Codenvy
  URL 152
config.xml file
  about 12
  structure 12, 13
continuous integration (CI) 294
controller
  creating 58, 59
  Jasmine tests, writing 280-282
Cordova application
  Crosswalk, adding 199, 200
  generating 150
  installation 9
  plugin, references 252
  URL 83
Cordova/Sencha Touch application
  authentication, implementing on
     application side 135-140
Cordova StatusBar plugin
  used, for fixing overlap 82-85
Crashlytics
  using 325-333
Crazy Bubbles game
  bubble position, detecting 163
  bubble, releasing 166
  bubbles, swapping 164, 165
  creating 153-157
  executing, on mobile 172
  game over screen, implementing 175

  game restart, implementing 184
  planning 149
  pointer, moving 162
  preparing 154-157
  score, calculating 172
  selected bubble, moving 163
  sharing, to Facebook 185-188
  sharing, to Instagram 189-191
  sharing, to other social media 185-188
  sharing, to Twitter 185-188
  sprite, displaying 158, 159
  sprite, preloading 157, 158
cross-platform workflow 8
Crosswalk
  about 198, 199
  adding, to Cordova application 199, 200
  benefits 199
current geolocation
  detecting 97, 98
custom plugin
  building 252
  JavaScript interface, adding 254
  native implementation 255, 256
  publishing 257
  setting up 253, 254
  using 257

D

DalekJS
  URL 286
  used, for testing 286-289
data
  displaying, with Google Maps 99-102
  picture details, displaying in
     popup 102-106
  saving, in local storage 99
development aspects, PhoneGap
  browser reflows, minimizing 39
  hardware acceleration, using 38
  images, optimizing 38
  network access, limiting 38
  payload, optimizing 39
  perceived speed, increasing 38
  single-page application approach, using 37
  testing 39
  UI generation, avoiding on server 37

development methods, PhoneGap
  cross-platform workflow 8
  platform-centered workflow 8
device profiles
  creating 64
Document Object Model (DOM) 146
dressed photo
  picture model, defining 248
  picture, saving to filesystem 249-251
  picture store, defining 248
  saving, to applications folder 248

E

effects
  applying, to photo 246, 247
  applying, to thumbnails 242-245
effects list
  effects model, defining 241
  effects store, defining 241
  photo popup, displaying 237-240
  Pixastic library, including 236, 237
  rendering 236
Express application, REST API
  exploring 117, 118
Ext JS
  URL 56

F

Fabric
  URL 326, 332
  using 325-333
Facebook
  Crazy Bubbles game, sharing 185-188
Famo.us 43
File API
  on W3C, URL 94
filesystem plugin
  installation 94
  persistent file location, using 94-97
  usage 94
file upload
  implementing, on application side 142-144
  implementing, on service side 140, 141
Flatly theme
  URL 127

Flixel
  URL 148
FontAwesome
  URL 39
Framework7 42
functional testing 270

G

game framework
  selecting 146
game over screen, Crazy Bubbles game
  horizontal scenario 177, 178
  implementing 175
  logic, coding 178-182
  vertical scenario 176, 177
GapDebug
  Android debugging setup 71
  Genymotion Android emulator 75
  installation 70
  iOS debugging setup 70
  OS and configuration requirements 70
  URL 70
  usage 70
Genymotion Android emulator
  about 75
  URL 75
Geolocation API
  URL 97
Git client
  about 10
  URL 10
Google Maps
  data, displaying 99-101
  icon parameter 101
  map parameter 101
  picture details, displaying
        in popup 102-106
  picture parameter 101
  position parameter 101
  title parameter 101
Google Play
  application, building in release
        mode 322, 323
  application, releasing 322
  application, signing in release
        mode 322, 323

application, uploading 324
keystore file, creating 322
URL 324

H

HAX
  installation and configuration, URL 75
headless browser PhantomJS
  used, for testing 272, 284, 285
Homebrew
  about 4
  installing 4, 5
  Node.js, installing 4, 5
  URL 4
HTML5 Canvas
  about 146
  example 147, 148
HTTP methods
  DELETE 108
  GET 108
  PATCH 108
  PUT 108
http-server, Node.js
  URL 152

I

Imaginary application
  Jasmine tests, writing 279, 280
IndexedDB 60
initial application MVC structure
  about 75
  controllers, organizing 81
  model 81
  Pictos icons, adding 79-81
  store 81
  views, organizing 76-79
Instagram
  Crazy Bubbles game, sharing 189-191
  plugin, URL 189
integration testing
  about 270
  with Jasmine 277
Ionic 41
iOS
  application, running in iOS emulator 16, 17

developer certificate, generating 19, 20
  setting up, for application 14, 15
  URL, for Human Interface Guidelines 39
iOS application, beta release
  distribution provisioning profile,
    generating 307, 308
  external testers, inviting 317-319
  internal testers, inviting 317-319
  performing 306
  uploading, to iTunes Connect with
    Application Loader 312-316
  uploading, to iTunes Connect with
    Xcode 309-311
iOS debugging setup, GapDebug
  about 70
  computer configuration 70
  iOS device configuration 71
iOS library
  custom plugin, building 252
items property 77
iTunes Connect
  iOS application, uploading with
    Application Loader 312-316
  iOS application, uploading with
    Xcode 309-311
  URL 306

J

Jasmine
  about 272
  URL 278
  used, for integration testing 277
  used, for testing 272
  used, for unit testing 273-276
Jasmine tests
  writing, for controller 280-282
  writing, for Imaginary application 278-280
  writing, for Sencha model 282, 283
Java Runtime Environment
  URL 47
JavaScript Object Notation (JSON) 110
JavaScript testing frameworks
  Jasmine 272
  JSSpec 272
  Mocha 272
  QUnit 272

JDK
  URL 28
jQuery Mobile 41
JWT
  URL 126

K

Kendo UI 42
Koding
  URL 152

L

launch process
  defining 65, 66
Linux
  Node.js, installing 6
local storage
  data, saving 99
LocalStorage 60
lorem ipsum text 273

M

Mac
  Node.js, installing from official
     website 2-4
  Node.js, installing with Homebrew 4, 5
MAMP
  URL 152
Microsoft IIS
  URL 152
mobile
  Crazy Bubbles game, executing 172
model
  defining 61, 62
MongoDB
  about 110
  and Express, connecting 121
  installing, with Homebrew 110

N

navigator.camera.getPicture function
  about 87
  DATA_URL parameter 87

  destinationType parameter 87
  quality parameter 87
Node.js
  about 2, 109
  advantages 109
  installing, on Linux 6
  installing, on Mac 2
  installing, on Windows 5
  URL 2
Node Package Manager (NPM)
  about 3
  PhoneGap, installing 7

O

Onsen UI 43
OpenTok
  about 225
  URL 225

P

PeerJS
  URL 216
  used, for building real-time communication
     application 214
performance testing
  about 270
  with Appium 289-293
  with browser-perf 289-293
PhantomJS 284
Phaser
  about 148-151
  downloading 151
  pointer events, handling 160, 161
  references 149
  text editor, using 151
  URL 151
  web server, using 151, 152
PhoneGap
  about 1, 7
  basic components 8
  Cordova, installation 9
  development methods 8
  installing, with NPM 7
PhoneGap Build
  config.xml file 300

creating 302-305
plugins, adding 301
uploading 302-305
URL 299
using 299
PhoneGap Developer App
about 263
code, modifying 266
core plugins, including 266, 267
setting up 264-266
URL 264
PhoneGap Social Sharing plugin
about 185
URL 185
PhoneRTC
about 225
URL 225
photos
capturing 234, 236
effects, applying 246, 247
listing 258-262
photo popup, displaying 237-240
Pictos font
URL 79
Pixastic library
including, for effects 236, 237
overview 228, 229
URL 228
Pixi.js
URL 148
platform-centered workflow 8
plugin registry
URL 88
plugins repository
URL 8
Postman Chrome plugin
URL 124
proxy
about 64
client proxies 64
server proxies 64
Pumpidu
about 194
URL 194

R

Ratchet 41
React 42
real-time communication application
building 200
building, with PeerJS 214
client side components, adding 202-210
Cordova application, tweaking 210
executing 210-214
server side components, adding 200-202
**real-time communication application,
    with PeerJS**
client side components, adding 216-220
executing 221-225
server side components, adding 214, 215
**REST Application Programming
    Interface (REST API)**
Base URI 108
basic Express application,
    exploring 117, 118
building, technologies 109
developing 111
discovering 108
Express and MongoDB, connecting 121, 122
Express application, generating 113-116
Express, using 111-113
HTTP methods 108
MongoDB, accessing 121, 122
picture model, creating 122-124
picture record, creating 125
picture record, editing 125
record, deleting 125
response, returning 120, 121
service authentication, implementing 126
URL path 108
URLs, handling with routes 119, 120
RTCPeerConnection API 197
Ruby
downloading 47
URL 47

S

SASS
URL 80
Sauce Labs 294

Sencha Cmd
  categories 48
  commands 48
  features 48
  installing 47, 48
  URL 47
Sencha Touch
  about 41, 46
  application, bootstrapping 230-233
  installing 47
  Sencha Cmd, installing 47, 48
  Sencha Touch SDK, installing 47
  URL 46
service authentication
  endpoint request, handling 129-132
  implementing 126, 127
  implementing, on application side 135-140
  login form, implementing 127, 128
  verifying 132-134
single-page application (SPA) 37
SIP.js 226
Socket.io client library
  URL 203
sprites
  references 38
store
  creating 62, 63
  using 60, 61
stress testing. *See* **performance testing**
Sublime Text
  URL 151
system testing 270

T

tabBarPosition property 77
technology requisites, REST API
  exploring 109
  MongoDB 110
  MongoDB, installing with Homebrew 110
  Node.js 109
Telerik Test Studio
  about 294
  setting up 294
  URL 294
test-driven development (TDD) 268, 269
testers
  about 317

  external testers 317
  internal testers 317
testing
  behavior-driven development (BDD) 269
  benefits 267
  test-driven development (TDD) 268, 269
  with DalekJS 286-289
  with headless browser
      PhantomJS 272, 284, 285
  with Jasmine 272
testing, classification
  functional testing 270
  integration testing 270
  performance testing 270
  system testing 270
  unit testing 270
testing tools
  about 293
  Sauce Labs 294
  Telerik Test Studio 294
test runners 271, 272
thumbnails
  effects, applying 242-245
tools
  for WebRTC mobile applications 225
  OpenTok 225
  PhoneRTC 225
Topcoat 42
Twitter
  Bootstrap, URL 127
  Crazy Bubbles game, sharing 185-188

U

UI framework
  Famo.us 43
  Framework7 42
  Ionic 41
  jQuery Mobile 41
  Kendo UI 42
  Onsen UI 43
  Ratchet 41
  React 42
  selecting 39, 40
  Sencha Touch 41
  Topcoat 42
unique device identifier (UDID) 22

unit testing
  about 270
  testing frameworks 271, 272
  with Jasmine 273-276
user experience (UX) 40
user interface (UI) 40

V

versioning 298
view
  defining 56-58

W

WAMP
  URL 152
Webkit 284
WebRTC
  about 194
  audio engine 195
  browser support 198
  protocol stack 196, 197
  reference link 226
  RTCPeerConnection API 197
  video engine 195
WebRTC, components
  MediaStream 194
  RTCDataChannel 194
  RTCPeerConnection 194
web server
  using 151, 152
Web SQL Database 60
Windows
  Node.js, installing 5

X

XAMPP
  URL 152
Xcode
  iOS application, uploading to iTunes
      Connect 309-311
  URL 14

Thank you for buying
PhoneGap By Example

About Packt Publishing

Packt, pronounced 'packed', published its first book, *Mastering phpMyAdmin for Effective MySQL Management*, in April 2004, and subsequently continued to specialize in publishing highly focused books on specific technologies and solutions.

Our books and publications share the experiences of your fellow IT professionals in adapting and customizing today's systems, applications, and frameworks. Our solution-based books give you the knowledge and power to customize the software and technologies you're using to get the job done. Packt books are more specific and less general than the IT books you have seen in the past. Our unique business model allows us to bring you more focused information, giving you more of what you need to know, and less of what you don't.

Packt is a modern yet unique publishing company that focuses on producing quality, cutting-edge books for communities of developers, administrators, and newbies alike. For more information, please visit our website at www.packtpub.com.

About Packt Open Source

In 2010, Packt launched two new brands, Packt Open Source and Packt Enterprise, in order to continue its focus on specialization. This book is part of the Packt Open Source brand, home to books published on software built around open source licenses, and offering information to anybody from advanced developers to budding web designers. The Open Source brand also runs Packt's Open Source Royalty Scheme, by which Packt gives a royalty to each open source project about whose software a book is sold.

Writing for Packt

We welcome all inquiries from people who are interested in authoring. Book proposals should be sent to author@packtpub.com. If your book idea is still at an early stage and you would like to discuss it first before writing a formal book proposal, then please contact us; one of our commissioning editors will get in touch with you.

We're not just looking for published authors; if you have strong technical skills but no writing experience, our experienced editors can help you develop a writing career, or simply get some additional reward for your expertise.

PhoneGap 3 Beginner's Guide

ISBN: 978-1-78216-098-4   Paperback: 308 pages

A guide to building cross-platform apps using the W3C standards-based Cordova/PhoneGap framework

1. Understand the fundamentals of cross-platform mobile application development from build to distribution.

2. Learn to implement the most common features of modern mobile applications.

3. Take advantage of native mobile device capabilities—including the camera, geolocation, and local storage—using HTML, CSS, and JavaScript.

PhoneGap Mobile Application Development Cookbook

ISBN: 978-1-84951-858-1   Paperback: 320 pages

Over 40 recipes to create mobile applications using the PhoneGap API with examples and clear instructions

1. Use the PhoneGap API to create native mobile applications that work on a wide range of mobile devices.

2. Discover the native device features and functions you can access and include within your applications.

3. Packed with clear and concise examples to show you how to easily build native mobile applications.

Please check **www.PacktPub.com** for information on our titles

Instant PhoneGap

ISBN: 978-1-78216-869-0   Paperback: 64 pages

Begin your journey of building amazing mobile applications using PhoneGap with the geolocation API

1. Learn something new in an Instant! A short, fast, focused guide delivering immediate results.

2. Build your first app using the geolocation API, reading the XML file, and PhoneGap.

3. Full code provided along with illustrations, images, and Cascading style sheets.

4. Develop an application in PhoneGap and submit it to app stores for different platforms.

PhoneGap 3.x Mobile Application Development HOTSHOT

ISBN: 978-1-78328-792-5   Paperback: 450 pages

Create useful and exciting real-world apps for iOS and Android devices with 12 fantastic projects

1. Use PhoneGap 3.x effectively to build real, functional mobile apps ranging from productivity apps to a simple arcade game.

2. Explore often-used design patterns in apps designed for mobile devices.

3. Fully practical, project-based approach to give you the confidence in developing your app independently.

Please check **www.PacktPub.com** for information on our titles